RONALD CICUREL

Spyridon's Quest

A fantastic novel in 3 parts illustrated by the author

SARINA PUBLISHING

SARINA PUBLISHING

Typeset in Baskerville
Printed in Switzerland

Translated from the original French
© Copyright July 2003 Ronald Cicurel
cicurel@bluewin.ch

... But if science becomes a subject of general interest, our perspectives about the world, knowing what it is and how to improve it and to improve ourselves will grow. Carl Sagan.

This is the story of Spyridon. Of his perseverance and courage. Of the reasons that have led him on long journeys through the past and future. Of his discussions with men of passion.
Men who have built up the history of human thinking. Real men, imaginary encounters. This is a science fiction story, mixing scientific facts with fiction. Although the facts are simplified and made easily accessible for the purpose of the story, the author hopes he has conveyed the full depth of the ideas.
He extends his thanks to the personalities whose names he has borrowed and hopes he has not put words into their mouths that they would not have said themselves...

Spyridon

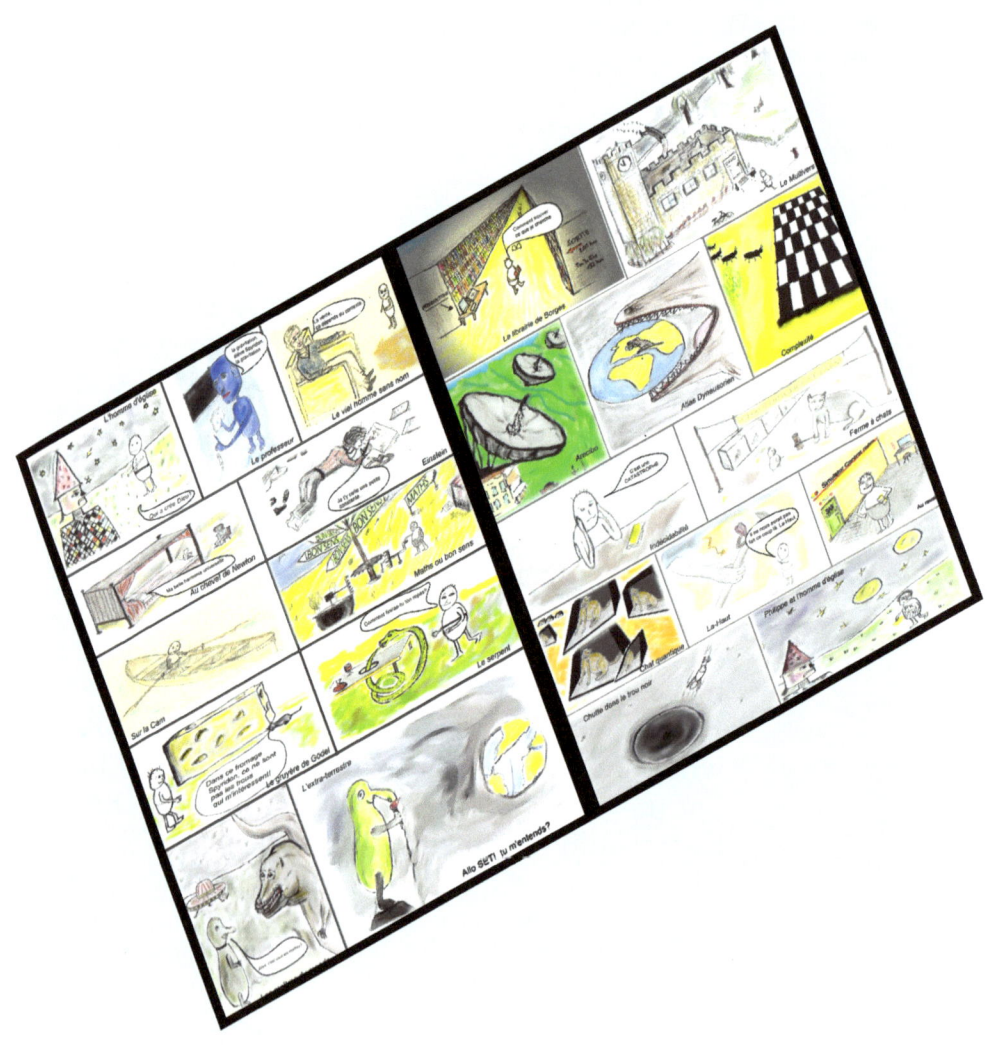

CONTENTS

FOREWORD   7

PART ONE: THE RISE OF CURIOSITY

1	WHO CREATED GOD?	10
2	WHAT MAKES APPLES FALL	12
3	WHO CREATED MATHS?	15
4	ENCOUNTER WITH NEWTON	21
5	WHO IS SPYRIDON?	23
6	ALBERT EINSTEIN WANTS A STABLE UNIVERSE	25
7	BLACK HOLES WITH STEPHEN HAWKING	29
8	ABOUT BARLEY BEER AND ENTROPY	33

PART TWO: KURT GÖDEL

9	WHERE THINKING STOPS	37
10	THE END OF ILLUSIONS	44
11	SPYRIDON MAKES THE POINT	50
12	IT'S SO BIG, THE UNIVERSE	55
13	THE TWO PHYSICS	60
14	MEN, DINOSAURS AND SHARKS	64

PART THREE: UNDERSTANDING

15	THE BEAUTIFUL SOPHIE	69
16	PARALLEL UNIVERSES	76
17	A LIBRARY WHERE YOU CAN FIND ALL PRESENT AND FUTURE BOOKS	83
18	THE COMPLEXITY INSTITUTE	87
19	ITS ONLY A GAME	95
20	EVERYTHING IS INFORMATION	100
21	SPYRIDON'S PLAN	103
22	THE QUANTUM COMPUTER	106
23	READY FOR THE BIG DEPARTURE	109
24	SPYRIDON DIVES INTO THE BLACK HOLE	113
25	THE END	117

SHORT GUIDE TO SCIENCE AND FICTION	119
IN ODER OF APPEARANCE	126
GLOSSARY	127

To my children and to she who has understood everything

To all the students of the world

WITH MY GRATEFUL ACKNOWLEDGEMENT TO:
Georges Badin
Liliane Mancassola

The authors who inspired me, in particular: David Deutsch, Lee Smolin, Murray Gell-Man...

In homage to Kurt Gödel

FOREWORD

Spyridon wants to "understand".

Spyridon is looking for answers to the questions we have all asked at some time or other: Who created the world? Where do we come from? Are we alone in the universe? Is there another reality behind this one?...

This is the story of his adventures: his encounters with famous men and the incredible plan he will build to obtain some answers. "Spyridon's Quest" is like a children's tale for adults.

During his journey, Spyridon learns, doubts, thinks and makes notes in his little yellow notepad. He discovers surprising and incredible things, true things. Spyridon does not accept ready-made answers or simplified solutions. He is critical towards man's ego and sceptical of any solution that would give a special place to humanity in the universe. He is, without knowing it, a defender of "Ockham's razor", the principle of parsimony: do not make any hypothesis that you do not need. He fears unconscious premises that guide us without our knowing it. Spyridon believes his adventure has a meaning. He will discover one that he could never have imagined.

Spyridon's quest, in search of the mystery of the creation, develops around his discovery of three important scientific concepts: self-reference and the incompleteness of arithmetic, the interpretation of quantum physics, quantum computers and simulations. The conjunction of these elements will lead him to think that the universe is part of a multiverse and that it behaves like a gigantic computer.

The black holes are going to play a determining role as an exit, which Spyridon wants to use.

I offer the reader the pleasure of being guided by Spyridon in this journey, which I wanted to create as a quest of modern times. In place of the usual physical tests in these types of stories, the obstacles, which Spyridon meets, are conceptual, but their confrontation requires the same courage and the same stamina.

From his meeting with the churchman, to his plunge into the black hole Cyggnus 2, Spyridon displays a naïvety and stubbornness, he remains a little boy, a cartoon character, trying to understand the world.

This story will certainly leave you hungry with so many questions remaining unanswered, but I trust you have an enormous appetite.

SPYRIDON'S QUEST

Ronald Cicurel

Part one : The Rise of Curiosity

1
WHO CREATED GOD?

Spyridon is still very young. He is at an age when human beings and cartoon characters are still not convinced they "know". He is naturally curious. So he asks questions. But Spyridon does not necessarily accept the answers one gives to him, because he wants to understand. Spyridon wants things to seem logical. At the beginning of this story, Spyridon is still at the age when he believes that adults do know....

Who created God?

- But…. If God created the world, who created God? Spyridon asked the churchman.

- He created himself, my son. He is the unique and almighty God.

- So why don't we say that the world has created itself? Why believe in a God?

- Because the scriptures tell us that God created the world.

- So one can create oneself, if one does not exist?

- Sure, if one is God.

- But if one is not yet God?

- That does not matter, as one is going to become God.

- So God exists before he creates himself? So why does he need to create himself?

- Spyridon, my son, you have faith already in your heart, it is your faith asking those questions. But it needs to grow and mature. Search Spyridon, searching is your way of loving...

I cannot be very logic thinks Spyridon, everybody else seem to understand and find things natural, and yet I don't understand anything. Funny...

I must start at the beginning, I want to know who created us, and why we are here.

2
WHAT MAKES APPLES FALL

Time has passed, Spyridon is already 14 years old, he has thought a lot about the answers of the priest and read the scriptures. But he is getting nowhere. Now at school, he questions his professor about causes and effects.

The teacher

- Teacher, why do apples fall and planets revolve?
- Well, because of Newton's laws, gravitation, Spyridon, you should know that.
- But what is gravitation?
- It's what makes one mass attract another.
- But why do they attract one another?
- Well, because of gravitation.

"I cannot be very logical… They all seem to understand and I don't. The answers I am getting go round in circles.

A is because of B and
B is because of A.

That cannot be an explanation, can it?

Every time I ask "Why" I am told "How" with the use of some equations. Is that understanding? I am not satisfied, I don't find this right".

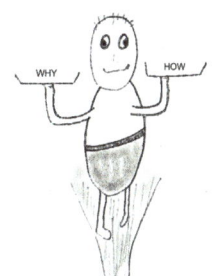

So Spyridon is disillusioned by the answers
he receives. First, it seems to him, that for
every effect there are one thousand causes,
naturally, some are more important. But all the
past is a part of causes. All this is "not very
clear". Spyridon needs help. He wants to be able
to speak to someone who also finds things "not very
clear". Unfortunetly things seem clear for everybody.
Spyridon feels very lonely. For example these equations
that he was taught at school, are they really the way to
understand? He is not very convinced of it. Spyridon
plunges into his textbooks, where he finds magnificent
things but not what he is trying to understand. Fortu-
nately close by there lives a wise old and nameless
man who has the reputation of not believing
anything easily. They say that he used to be
a mathematician. Spyridon wants to have his
opinion on mathematics and its usefulness. He
wants to know if he is the only one asking
these questions.
Is mathematics so essential to our "understanding", or could we do without it? Where
does it come from? Many times he stopped
in front of the old man's door. Then one
morning, he decides to knock.

The old nameless man

3
WHO CREATED MATHEMATICS ?

Where Spyridon consults the wise man with no name who does not accept ready-made answers.

The man who opens the door has the very particular look of those who live more inside themselves than in the outside world. Spyridon likes him immediately and his fears dissolve.

- Mister wise man, you have the reputation of not believing anything without proof.

- It has not made my life easy. You can see it, Spyridon, I am old and poor. That is what happens to people who do not conform to established rules. And I had the obsession for truth, not for rules.

- How do you distinguish between what is "true" and what is "false"?

- Precisely, in most cases it is very difficult, it is never final, umm..., each one has his definition, umm...., let's say it depends on the context. In fact it's one of the most difficult problems. Plato thought that there is a world of ideas that is much more "real" that the one our eyes perceive. Absolute truths of which we are more or less aware. Many of us still believe Plato.

- Would you say for instance that mathematics is "true"?

- Yes, umm… for mathematicians, once you have accepted the starting points and the construction method. But there are limits, undecidable things. We can access the "false", but the "true" is never absolute.

- I just came to ask you, if you really need all this mathematics to understand the "realities" of the world?

-Your question is complicated... One can do without it in everyday life, naturally. One bases oneself on experience and intuition and this is quite enough in most cases. But certainly not enough to build planes or computers or to understand matter, time or space. Have you read Descartes?

-No.

-Some time ago, he wrote a small book, which made an incredible impact: "The Discourse on the Method", he noticed that common sense is one of the best shared things and that everyone claims he has enough of it.

-So what?

-Common sense is not enough, because there are different common senses according to each person. If one wants "a common knowledge" one needs a method upon which each one of us can agree and recognize himself. A method and a language.

-Is this therefore the reason that mathematics is necessary?

-When one wants to describe the very small or the very big, for example, or to be more precise with a forecast, it is necessary to be able to rely on a language, which is better adapted to what one is describing and whose rules are known by everybody. With such a language, one can produce more refined and more appropriate models. Then everyone who has studied enough can accept them. But our knowledge is far broader than what we can learn in this way. One should not confuse the language with what it expresses.

-So mathematics is a language?

- Yes a language and other things. A language and a guide. Leibniz was dreaming of a language with which you could only express truth, like Ariadne's thread of the though.

- But who established the rules? Is mathematics a product of nature or was it invented by mathematicians?

- Since Plato, many schools of thought have developed. Mathematics satisfies the main criteria of objectivity: It remains the same whoever studies it. Your question is far-reaching. I have not been so far. On the one hand, I believe that it is universal, it is part of nature, of the universe. Extraterrestrial beings would then have the same mathematics as we have. On the other hand, I believe we have invented it. The discussion goes way back to Plato and Aristotle. Plato would say that mathematics is "real" and fundamental. Aristotle would claim that it is physical reality, which is fundamental, and that mathematics is only a language that describes reality.

- Ah! Plato would say "discovering", like you discover an unknown land that already existed before and Aristotle would say "inventing", like you compose new music with notes, which had never before been arranged in that way.

- Yes, and it makes a big difference with huge consequences concerning the way we understand reality. For instance, your concept of "existence" differs greatly, depending on whether you think we "discover" or "invent". The same thing goes for concepts like "objectivity", "reality", "love",... People consider mathematics as a form of truth on which any mathematician can agree. Though they also have their limits.

- What limits ?

- Those undecidable things. You should meet Kurt Gödel, he will tell you more than I can.

- Who is this Kurt ?

- In 1900 during the mathematical congress the famous mathematician David Hilbert published a list of 23 problems he believed would be solved during the 20th century. His second problem was more or less to prove that arithmetic contains no contradictions. Gödel is the man who solved it. As you can see this is quite close to your questions.

- Seams obvious, but OK. I will go to meet him.
- Today mathematics has become the priveliged language of most sciences, from the most fundamental such as physics, which uses it a lot, to the higher level ones like biology. A theory is often a mathematical model. A sort of virtual image of reality written in maths. So mathematics plays a special role in the sciences.
A bit outside, above and to the side. The only test of reality to which it is submitted is the test of consistency, whereas other sciences undergo the test of real experience.
- Why have you got no name on your door?
- Nobody comes to see me… Well, those who want, find me. Then I want no confusion with that...
- You don't want to be confused with yourself?
- Things are not what you name them..., I am not "me".

Spyridon came out of this meeting very satisfied :
" I truly believe that we need a language and that mathematics helps to prove theories and make models from them and to communicate, but discovering and understanding is something else, there are other dimensions."
"One understands first and proves things afterwards..."

He wrote this little sentence on the first page of his yellow notebook. He also added the name of Kurt Gödel and promised himself he'd go to meet him soon.
The wise man had placed mathematics at the top of the science ladder, but a little apart from all the others.
Suddenly Spyridon has an idea:
"Mathematics is the result of a process of our mind, it is necessarily a result of the laws of physics. It originates through physical processes and so is part of physics".

Maths flowers in the nature

This thought seems very strange and in any case very far from Plato. "Either it is part of physics, or it is outside and serves as a tool to develop physics". Spyridon thought he had found an answer, but now here he is with new doubts and new questions! The encounter with the wise man had comforted him in his conviction: **"The only thing I am interested in is knowledge, I will devote my time and my life to it"**.

Spyridon's decison was taken, he would travel, meet all the people he would need to, in order to arrive at his own ideas on subjects he is interested in. He would travel in the past and in the future.

Travels reserved for cartoon character obviously. But Spyridon is not an ordinary child, he is a cartoon charater, which makes him a little bit flat and naive, but it is a huge advantage for his daily routine, the exercise of his imagination and his ease of travel.

A simple little drawing in his magic yellow notebook and he can be wherever and whenever he wants on Earth.

Before his first trip, Spyridon, inspired by Hilbert, establishes a list of questions to resolve:

1. Why are we here?
2. Who made us?
3. Who made the world?
4. Was there a beginning? And before this beginning?
5. Can one know the laws which govern the world?
6. Are there extraterrestrials? How can we meet them?
7. Can one meet God?
8. Why does it seem to me that I do not understand what others appear to understand?
9. How does one recognize the truth?
10. Is mathematics a branch of physics?

Every time he finds a totally acceptable answer for himself, he will cross off the question. He will then establish a new list. He already knows that no answer will be definitive and that new questions will arise on his journeys. But at least he has an embryo of a method.

Spyridon's first trip will be to meet Newton, the person who explained why apples fall and planets revolve. Newton is universally recognized as one of the greatest geniuses of all mankind. During the outbreak of the plague, which raged in Cambridge in 1665 and 1666, he withdrew to the family estate, where it is said that he had most of his important revelations: universal gravitation, laws of motion, calculus, decomposition of white light. That year was called the "annus mirabilis" by the historians. But before leaving, Spyridon goes to the nearby library, where he reads the "Discourse on the Method" …

"Common sense is the best shared thing in the world, because each one of us thinks he has so much of it, …"

4
ENCOUNTER WITH NEWTON

Spyridon is on his first cartoon trip. He has taken his yellow booklet, and made his first drawing. Here he is in 1727, in London. He is sitting on the bed of the famous Isaac Newton. The man who has explained why apples fall and planets revolve. Newton is old and sick. Behind the bed, Halley the famous astronomer, the one who discovered the comet, is sitting on a chair.

- Tell me, Halley, who is this character sitting on my bed?

- Come on Newton, it's Spyridon, he has come from the 21st century.

- Spyridon, says Newton, am I famous in the 21st century? And my equations, my three Laws… and my calculus. Yes, my calculus is it still used?

- Yes, Mister Newton, you are very famous and your calculus is taught in all the schools.

- Do my equations still inspire humanity with their splendid harmony?

- Umm! yes, but... they are a little bit out-dated.

- How is that possible. But at least has this Leibniz been forgotten?

- No, not quite Mister Newton. We where speaking about him not long ago.

- Ah! Ah! It makes me sick…

- ...

- Tell everybody that I am the one who invented differential calculus, this Leibniz is only a copier, an imposter, a fake!! I cannot understand that the marvellous universal harmony that I bequeathed you can be outdated.
- Why do apples fall, Mister Newton?
- You have not read my work, young man!
- Yes, Yes, Mister Newton, but what I want to know is "why" they fall, not "how" they fall.
- Gravitation young man, Gravitation.

Spyridon concludes that Newton's lesson had been faithfully handed down from generation to generation and that his teacher had served him the same stuff…

"It is not because something is old that it is more correct, errors go through generations at least as well as truth". He has not learned much from the great Newton, probably he arrived at the wrong moment. Probably he was expecting too much, he had allowed his hopes to rise too high.
"I did not expect this. I believed that great thinkers would also be great men. Better not to expect anything next time and simply ask my questions, so I will not be disappointed. I will be more open to perceive reality rather than my own expectations of it". "I will also have to consider some strange human aspects: pride, jealousy, slander…" he said to himself.

5
WHO IS SPYRIDON?

My name is Spyridon. I now live where my questions take me. Sometimes I travel in time. My life is a sort of cartoon where I play the hero. I am sure about nothing, I want to learn. When I was younger, I did not understand the answers I was getting. I found myself stupid and did not know what to say, especially when all the others seamed to understand and accept what was being said.

Now I simply laugh about it.

Particularly about men.

Those who believe they are so important.

Those who put themselves at the center of everything.

Those who are bothered only about their petty daily problems.

Those who have forgotten the real questions and believe they know everything.

Those who act as if the whole universe has been built only for them.

Those who have a pain in the back from looking at their own navel too much.

Those who chase after the honours.

Those who are envious of others who are also envious of them.

Those who are blinded by themselves.

This pride makes me laugh.

I want to know why we are here, how all this has been made, and if behind the reality we see, there is a will, a meaning, a mission or a destiny.

I want it so badly that for me the rest is not important, everybody can want a beautiful car, wonderful holidays or a new fridge. I don't, I want to know, to know what knowing and understanding mean.

Understanding is fantastic, not only to accept other's answers but conceive one's own, reconsider one's judgment at the slightest new observation or unexpected comment. Understanding is what gives meaning to things, a reason for our existence.

I do not expect things to be easy but as I am young, I am in a hurry and impatient.

I feel free in my mind. I want to choose my own way, decide by myself what is important. I want to choose my own teachers. Make my own mistakes.

As a cartoon character, I am very lucky, I am free of many constraints, and I intend to take advantage of this situation.

I decided to start by speaking to men whom I imagine to have the same hunger as mine. Then I will make my plans. My next trip will be to meet the man who outdated Newton. I read that Albert Einstein was considered as the greatest physicist of the 20th century, not only because of relativity, but also because of his major contributions to quantum physics for which he received the Nobel price. It must be very nice man with such a face. What's more, he declined the presidency of Israel to pursue his work in physics. But I prefer meeting him when he was young and in his prime.

Spyridon quickly made a little drawing.

6
ALBERT EINSTEIN WANTS A STABLE UNIVERS

When Spyridon arrives at Einstein's house in 1915, Albert is putting the final touches to his general relativity theory. Papers are strewn all over the place, Albert had been working very hard on his theory since 1905, he had developed the ideas, but certain mathematical tools where lacking.

I will add in a little constant here.

- These equations don't work! They show a universe expanding or contracting. No, the universe must be eternal and infinite, I will add a little constant here to balance everything, it will do the job.

- And if it was true, Albert, that it is expanding?

- Impossible, Spyridon, it would then have a beginning and if it had a beginning there would be something before the beginning and my equations show nothing, they collapse when you get too close to the beginning. So no beginning, no end and no expansion. My little constant will do the job.

- Perhaps it's a stroke of genius, or perhaps you will regret it!

- What do you know? With no constant the whole universe comes crashing down, that is not what He wanted "up there"!

- Do you know "His" intentions…?

- You don't understand Spyridon. Better a little constant and a stable universe that has always existed, rather than a beginning and having to ask what went before.

- Your idea, Albert, to describe gravitation as a curvature of space-time and transform geometry into a branch of physics is genius.

- Yes, I know. The idea is correct. But it took so much time and Hilbert is also working at it. I am glad I was able to find the right mathematics. In the end I got help, but what a job from the moment I became aware of the Michelson-Morley experiment with the speed of light.

- Do you believe, Albert, that one will be able to describe total physical reality with a mathematical model?

- Yes, that is sure Spyridon. What funny ideas you have. I believe it. I want to believe it. He would not have "done that" to us, "up there". There is a logic to this world, and it is available to us. When I have finished with Relativity, I will set about this Unitarian theory, I will have to integrate Maxwell's electromagnetism...

He wants it that way, thought Spyridon: A universe with no beginning, no end, infinite and a creator that does not chance the dice, (this expression came to Einstein later when he was confronted with quantum indetermination).

"The most common things are the hardest to understand", wrote Spyridon in his yellow notebook, thinking of space and time. "The most dangerous hypotheses are those that we presume without knowing it. They influence our conclusions without us noticing".

"Einstein wants to avoid a beginning because his equations do not describe it".

Next to this phrase he wrote :

"Einstein believes that one can describe everything with equations".

He remembered a similar idea, which had occurred to him as he was leaving the wise man:

"Is Maths part of physics".

He compared this with:

"According to Einstein you should be able to write all physics with equations..." Funny...

He is slowly feeling an idea taking shape, something like:

"A is part of B which is part of A".

Spyridon sits down on a bench and looks at the sun. "The light I am seeing left the sun eight minutes ago," he thinks, "Einstein threw out the idea that two events could be simultaneous. Newton's notion of absolute time didn't exist any more. This means that we can only be aware of events, which have had the time to announce themselves to us".

The other question raised with Albert, about the beginning of time and the equations, which don't solve anything, was puzzling him enormously. Spyridon had heard about Stephen Hawking, professor at Cambridge University who, despite a paralysing illness, was recognized as one of the biggest geniuses of the 20th century. He seems sympathetic, observed Spyridon, and made another drawing in his booklet to move forward in time to 70 years later.

7
BLACK HOLES WITH STEPHEN HAWKING

We are in spring 1985. Stephen Hawking is in his apartment in Cambridge, a broad smile spreads over his face already deformed by his illness. His eyes sparkle behind the thick glasses, like those of a naughty child, about to make a bad joke. As soon as he perceives Spyridon, he exclaims:

-Come here, Spyridon, come here I cannot speak very loud, you know.

-I would like to ask some questions...

-About the black holes?

-???

-You know, astronomical black holes, these are stars, which have used up all their nuclear fuel and cannot resist the pressure of gravitation any more. They have collapsed to the point where their density has become enormous, so that nothing, not even light, can escape any more, their gravitation is so strong. The black holes bend space to the point where it closes in on itself. That is the reason they are black. And I showed with Bekenstein that, because of quantum phenomena, black holes still shine.

-Oh!... Good!... Mister Hawking, I had come to ask you a simpler question, Einstein's equations, do they show a beginning to the universe?

-Yes, naturally, Spyridon, the "Big Bang", about 14 billion years ago. We have an excellent description of what happens after the first millionth of a second (more exactly 10^{-43} seconds), but before that, closer to the big bang, the original singularity, fundamental forces become grouped together, the equations collapse, infinities appear and they lose all meaning. To go further and to know what happened at the very beginning of the universe, we would need a "Unitarian theory", which combines all the fundamental forces. For the moment this does not exist.

-Oh! So nothing has changed then … How do you understand that the world is what it is and how do you explain that we are here?

-Ummm! The best that I can say to this, without making an act of faith, is the anthropic principle. Listen; if anything in this universe had been different from what we observe, we would not be there to ask the question why we are here, or why are things the way they are!

-What do you mean?

-If one of the universal constants, the initial conditions, as Planck's constant, the speed of light, the constant of gravitation had been even just a little different, galaxies would not have formed, stars would not have burned, the earth would not have emerged, and we wouldn't be here either. So we would have no questions!

-You mean that the universe came into being in just such a way that would prompt us to ask the question: "why are we there?"

-Yes, that is exactly it, there is no other possibility, it is useless to ask the question why!

-You mean it is useless to ask the questions, which the universe's very evolution prompts us to ask!! Is that it?

-Yes.

-Are you sure, Mister Hawking, that it is not a simple word game this "anthropic principle"?

-No, no, think about it. It is a real physical principle.

Cambridge is particularly beautiful and surprisingly warm under the spring sun. Spyridon rents a boat to go out on the river Cam and to think about Stephen's comments.

"Curious, curious. What Hawking said to me sounds a little like the answer that the churchman gave me a long time ago, Spyridon says to himself.

This anthropic principle makes me wonder. Something is strange here. It is not right. That is no explanation. As for the part about the beginning of time, nothing new again, one comes quite close, but one does not actually get right to it!"

Settled in the boat, Spyridon opens his yellow pad and consults the list of 10 questions which he had drafted originally. So far, none could be really crossed off. With the benefit of hindsight, Spyridon finds his list a little bit naive and ridiculous. His questions are too general and to move forward he would have to go into more detail. If he had to rewrite it today, he would do it differently. It is not so much the questions that he finds ridiculous, it is the approach: the belief that it is possible to ask questions and to receive answers and be satisfied.

"Knowledge is a more complex exercise than my small list of questions with attached answers. There are so many forms of knowledge, so many different criteria for selecting answers".

"How can one be sure that an answer is true?"

"I met the greatest geniuses Newton, Einstein, Hawking, and the simplest things such as space and time are more and more mysterious to me... I read Descartes and I am not yet convinced of what is true and what is false... The churchman did not say anything to help me understand things better today. And although we are supposed to know so many things, we actually don't know anything."

Where should one begin? He moors the boat in front of Kings College and wanders up to the library. Very soon he stumbles across James Clerc Maxwell's name, the father of electromagnetism and one of greatest physicists of 19th century. Einstein had mentioned him when speaking about the Unitarian theory, he wanted to integrate Maxwell. Doubtless to make his Unitarian theory he needs to combine electromagnetism with general relativity. Spyridon has the feeling that this is the man he must now meet.

To find him, he needs to step back more than 100 years into the past. For him this is not too complicated. A little pencil tracing, representing Maxwell in front of a fresh beer (Spyridon could enjoy one too, in this heat) and off to go.

8
ABOUT BARLEY BEER AND ENTROPY

This winter in 1874, Maxwell is sitting in a sort of laboratory in Cambridge, the Cavendish laboratory, which he designed himself. The tables are dotted with glasses of beer, it was one year ago that he first published his four equations describing the behaviour of the electromagnetic field. This field which Einstein will later want to integrate. His wife stayed in Scotland and Maxwell feels a little bit sad.

-Wasn't it you who wrote that nothing gets lost and nothing is gained, asks Spyridon by way of an introduction.

-Hello Spyridon, you'll have a glass of a barley beer?... umm, no, you're getting mixed up... It was this French guy, Sadi Carnot, the man with the Carnot machine and the thermodynamics.

-What does Carnot say?

-In a system isolated from the rest of the world, the total energy remains constant. It is a bit like what the other French guy said, you know, Descartes. Today, it is known as the first principle of thermodynamics. But he also expressed a second one which says, roughly, that the quality of the energy of an isolated system degrades in time and that is interesting, especially if you add a little devil to it, like I did.

-What do you mean?

-It's as if you run a film backwards, let us say that you filmed your glass of barley beer falling and breaking on the ground. (You will be able to do this, once you have invented cinema). In the reverse sense, you would see the barley beer pouring back, the glass fragments sticking back together, and the whole thing jumping back onto the table. You would know at once that it is a film, in reality this does not happen, or at least it is very improbable. Virtual realities, such as the cinema, are not subject to the same limitations as our reality!.

-Yes, certain things happen only in one direction.

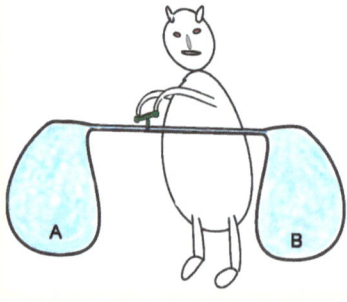

Maxwell's devil manipulates the system's tap so that the warm molecules group themselves in one of the containers and the cold ones in the other one, which appears at first sight to be contrary to the second principle... Szilard showed however that the process of measurement would consume more energy than the system would create.

-That's it, there is a direction in time. A time arrow. You should not be here because you come of 21st century.

-Nevertheless, I am here!

-It is because you are a cartoon caracter... I knew it straight away from your appearance. You are a "virtual reality". Imagination travels in time. In theory you could, for example, kill me. This would change the fate of the universe completely. And if you killed your own great-grandfather before he meets your great-grandmother, then you would not be there to kill him... You could also teach me some techniques from your century, of which I would become the inventor and which would then have no true inventor any more. You see it needs cartoons to be able to do what you are doing. Or a "good virtual reality" with its own laws of physics.

-What do you mean, when you talk about the degradation of the quality of energy?

-In fact, I do not totally understand myself, but it's as if the information disappears. There is more information in your glass of barley beer before it falls and breaks than afterwards. There is more disorder after the fall... this is entropy. And the second principle asserted in the proposition of Bolztmann says that the entropy of an isolated system is always increasing. That is to say that the disorder of the system increases with time. Energy, although preserved, is "less useful". I imagined a devil who by collecting and by handling information, fights against entropy and creates a machine with perpetual motion. You see energy and information are interlinked. It's as if by containing less information energy loses in quality.

-Entropy is therefore a measure of disorder.

-Yes, but also of information: the number of "yes" or "no" answers that this isolated system can provide.

-Then it is information, which is dissipating...

-Yes. We cannot create matter or energy, only transform it "by handling" information. When we say that we create, that is what we do, we list, select, assemble and reorganize information to attribute a meaning, it is our only method of creation. "To understand", is to establish the mental plan which allows us this manipulation. And when it happens, we feel as a sort of click, a miracle: EUREKA.

The meeting with Maxwell is immensely profitable. Spyridon eventually gets used to the barley beer, breaks a number of glasses, but has the feeling that he is understanding something very important. It is not yet totally clear what, but he notes some ideas in his yellow notebook:
"The universe handles information somehow or other, as would a huge computer. When the glass falls and breaks, the information it contains is handled, reorganized and partially lost, the universe changes state."

He notes, still not understanding why:
"The universe is like a huge computer which all the time is calculating its state from each moment to the next. Matter and energy are not only the result of this calculation but also the very calculation itself."
A sentence comes to his mind a little later and he notes it on a special page:

I am a part of the universe.

It is self-evident, but he feels he is on the brink of a big discovery. He, Spyridon, is one element used by the universe to make its gigantic calculation.
"I am one of the elements, which attribute a meaning to the information."

Part two : Kurt Gödel

9

WHERE THINKING STOPS

Spyridon has now become an adult cartoon character. Time has passed. He is now lying in a hammock, looking at the stars and thinking. He thinks about how we acquire knowledge, about information, about entropy, about the relationship between information, entropy, energy, meaning and all these concepts which he has discovered.

Spyridon on his hammock looking at the stars

"Unless I believe in Plato, it seems that a material basis is always needed to support information, particles or energy in certain states. Maybe matter and energy are only an illusion of our mind to capture the streams of information, as is the case with the colour of an object. Out of this information we produce something rather magic, we somehow attribute a meaning that allows us to understand.
I will have to think about it."
"I am beginning to feel the meaning of the word "to understand": it is something like selecting information, interpreting it, organizing it according to a model, a logic, a "preconceived" system connected with the other systems that we have already understood. And when it occurs, we attribute a meaning and we feel a sort of elation as if a miracle had occurred.
This miracle comes from the perfect compatibility between what was preconceived and the collected information.
One understands first, one proves later.
All information is available in nature, we have the incredible talent to select it, to organize it, to connect it to other information in a way that the whole becomes compatible and meaning emerges.
That is why we can understand in a false or limited way. If it was not possible no understanding would be possible.
By understanding I am a part of the job of the universe which is accomplished through me".
"But are there other ways of understanding, such as when one understands music or a poem? And then, understanding is not yet knowing... What is the difference between scientific knowledge and the other forms of knowledge?"

"As far as the questions about the passage of time, the beginning of time and God the creator are concerned… I have not advanced a single step since talking with the churchman. There are so many barriers to knowing."

Spyridon feels daunted by the work ahead of him. To start with he tries to make a list of these barriers he has encountered. He hopes to understand some hidden mechanisms and how the contradictions have come about.

- There is first of all the thing about the "beginning of time" and what happens "before the beginning". It may just be a simple trap of the language: if one speaks about the beginning of a system, there is no "before this beginning" within the system; but, as we have the habit of observing things from outside, we see when they begin and when they finish. Speaking about the universe, as a whole, we cannot observe it from the outside.

- There is the problem of "killing one's great grandfather" Maxwell spoke about it. It also has to do with the direction of time, one kills the cause with the effect…

- Then there is the question of the anthropic principle. Hawking insists that it is not a play on words. It excludes all other possible universes, because we are in this one, it is terribly logical, however I remain very suspicious. I don't think that the anthropic principle really answers the question "why are things the way they are". It precludes the possibility of asking the question what if things had been different. Here again, it depends on the fact that we are inside the universe.

- There is the paradox of the liar: a liar who says "I am lying", is not lying and yet, since he is speaking the truth, he is lying.

- There are explanations such as the one of the professor about causes and effects. A is because of B and B is because of A.

The snake

Spyridon's thoughts wander about in this way, without really arriving at any conclusions. He feels vaguely that there is something for him to discover in these paradoxes. And, as he is a cartoon character, he can take all the time he needs to reflect undisturbed, either while eating, or while looking at the match on TV, or while having his hair cut.

While he is thinking, suddenly a picture comes into his mind:

"A snake biting its own tail".

"That's it, precisely,", he says to himself, "How is the snake going to finish its meal?"

"The problem is that he is both the consumer and the meal … Yes, that's it, it can work as long as he stays far enough from the limits. It is like the equations of Einstein, which don't apply any more to the beginning of the universe. It is both the transformation and the object to which this transformation applies. It is in the same situation as God who "created himself", it is the creator and the object of the creation. This works, but only far from the limits, which means far from creation of itself."

"It's the same for the liar, after all, when he says "I am lying", everything is OK, as long as one does not apply the sentence to oneself."

"As for the anthropic principle, it is a little more complicated, but similar, it seems to me. The very existence of the question is used as an answer to the question."

"It is when one arrives at the limits that everything gets complicated. In the beginning the snake feels at ease, but when it approaches the end of its meal, what happens?"

Spyridon makes notes.

He feels as if he's bursting with enjoyment, his brain is working at top speed:

"All my examples make some use or other of an object, which has two roles at the same moment, one referring to the other". "One thing" which refers to itself and which is considered on two levels. The consumer is the meal. An "entity" that is acting upon itself. A part of the whole, which is acting upon the whole."

Carried away by his reflections, he notes in bold in his pad:

"When there is self-reference, things become confused".

Spyridon feels he has understood something of great significance and this inspires him enormously. He has succeeded in finding a common point among many paradoxical situations, in which logic ceases to apply any more.

"Understand... I am in the process of understanding!"

"Oh! But "understand" is also a self-reference word, how does one understand the word "understand", this is probably why I find it hard to define"...

This self-reference seems to be everywhere, a sort of property of our consciousness or maybe even of the universe. Hawking was no doubt right, this anthropic principle was not just a play on words.

And the more he thinks of it, the more he finds examples of self-referring systems such as: language, the word "I", the sentence "this proposition is unprovable", self-criticism... and every time it leads to paradoxical situations. However, what really lies underneath is still not completely clear to him. He notes:

"It's as if self-reference were introducing a limitation, a horizon to knowledge. A point where the spirit becomes confused. Is it actually the human mind or rather nature itself, which becomes confused?" He wondered. "No, nature knows exactly what it's doing, it is our brain which is limited".

That night he dreamed about a Swiss Gruyère cheese, knowledge riddled with holes where knowledge cannot go. When he wakes up, he has a revelation:

"Mathematics is in the same situation as the snake, it is both the builder and the construction. Mathematics is about mathematics, it is also a self-referring system".

"Yes", he says to himself, "but there is always an accepted starting point, axioms on which everything else is constructed". This thought reassured him a little.

"If there is any place where there cannot be any paradoxes, it has to be mathematics! However, I shall have to make some more enquiries, it would be a disaster. Who knows if my old nameless man was not thinking of this, when he said to me that even in mathematics there are limits..."

Spyridon spends a few more weeks meditating on his hammock. He wants to be fully prepared, before going to see Kurt Gödel, because he knows that this is now the man that he has to meet, if he wants to move forward with his quest.

The self-referring loop, which intrigues him most is the following one:

"Mathematics describes the whole of physical reality".

"Mathematics is a part of physics".

Theoretical physics would therefore also contain possible contradictions or limits …

Another intriguing loop is:
  "God has created the universe".
  "God is a part of the universe".
No, this does not hold water.
  "God has to be outside the universe," he says to himself,
  "He must have created it from the outside and could not have left himself much chance of intervening later on".

The moment arrived.
Having again researched on the Internet, Spyridon begins his journey, he remembers Maurits Cornelius Escher's drawing, which ought to lead him to the right place and draws it from memory in his notebook.

Kurt Gödel was apparently a strange person and became a close friend of Einstein towards the end of his life. They stayed together at Princeton and their walks from the Institute to their home became legendary. Einstein considered Gödel's company as a real honour. One day, Kurt announced to Albert that he had found a specific solution for the equations of general relativity corresponding with a universe, where time does not exist. Everywhere Kurt Gödel is considered as the greatest logician since Aristotle. From a young boy, he suffered from a paranoia that worsened with age. He finally starved to death in 1978, not eating any more, because he was convinced that his food was poisoned!

10
THE END OF ILLUSIONS

When Spyridon arrives in Vienna we are at the beginning of 1933, Hitler has just taken power in Germany. One of Gödel's professors, Moritz Schlick, was murdered by students of National Sozialism and Gödel is at the edge of his first nervous breakdown. In the salon, young Kurt is having lunch with a woman. Behind the thick, totally round glasses, his look is piercing, lively and anxious. The atmosphere in the apartment is heavy. The decoration is austere, almost Kafkaesque and the room seems to be taken from a black and white film. The woman tastes the food before serving it on the plates. Gödel is certainly not an easy person to live with. Spyridon addresses him timidly.

-I was advised to come and see you.

-Will you take some soup, prepared and pre-tasted!

-No thank you, Mr Gödel. I need to know if there are any limits in mathematics. Things that one cannot demonstrate.

-This is what I demonstrated two years ago, some undecidable propositions do exist. Mathematics is not complete, it contains propositions, which are true and yet cannot be demonstrated, contrary to what Hilbert, Russell and almost all the others thought.

-You mean propositions too difficult to demonstrate at the moment!

-No, no, I mean unprovable, by definition.

-It is not possible. So mathematics cannot prove itself completely. This is a disaster!

-If that's the way you feel. The truth is never a disaster. It's lying that is inadmissible! Listen to this Hitler!

-Have you really demonstrated, Kurt, that in mathematics, whatever system of axioms you choose, you will find true and unprovable propositions … How did you do it?

-I built such a proposition. My proof is water-tight. I took over one of Cantor's ideas, the diagonal argument, and there we are. In fact I found a way to code each proposition on numbers with a number. This system allowed me to arrange things as if arithmetic were about arithmetic. I was able therefore to build propositions, which are provable if, and only if, they are unprovable. Thus, I am able to translate the self-referring sentence such as: "this proposition is unprovable" into a proposition about numbers. I had to take a lot of care and precautions.

-.!!

-There is even more: not only a system containing arithmetic cannot prove its own consistency, but one cannot know in advance if a proposition of the system is provable or unprovable, until one has demonstrated it. There is no other criterion of proof than the very demonstration itself.

-Incredible … This is a disaster!!

-No, it is an old belief, which is crumbling. An unprovable proposition is often demonstrable in a more powerful system. You can for example add it to the axioms of the system. Then it is not necessarily so serious in practice, but, of course, the more powerful system in its turn contains its own unprovables... So you see, if one wants to limit oneself to a system based on a finite number of axioms... My theorem questions a very old paradigm.

-What is "true" is not the same as what is "provable"!

-Correct. It is necessary for us to revise concepts such as "truth", "provable", "exist".

-But then, physics also will always be incomplete!

-Yes it is very likely. First of all, a global physical theory, describing the totality of the universe, is a "mathematical model", but also such a theory is self-referring, indeed, if it describes everything, it should be able to describe itself in particular.

-But Einstein believes in completeness, because he is working on a unitarian theory, which would be a description of everything, of the entire universe. He said to me himself in 1915: "He would not have done this to us, up there". And I know that since then, he has been spending most of his time on this!

-And the Hitler phenomenon, "He, up there", did not hesitate... Einstein has probably still not read my theorems. Unless he has another idea in mind. Nobody had any real doubt about completeness, until my negative demonstration, it was just a specialist's detail to be proved, this is why they did not like my article. But finally, they had to go along with it. One cannot acquire all the knowledge of a system, which contains numbers, from inside the system.

-What do you mean, Kurt? In fact, I have already noticed that without being a mathematician.

-Everybody can see it. Take a dictionary as an example of a self-referring system, to define the first word, you are obliged to use words which will only be defined farther on. The dictionary remains useful, but the whole language cannot be defined by the language. Elements of accepted departure are needed, such as the axioms in mathematics. My theorem shows that whatever the selected axioms, the system cannot prove its own coherence.

-Yes, this is exactly what I noticed. There is as a logical breakdown in knowledge, as soon as a system refers to itself or is about itself. It means that we shall probably never be able to have a complete mathematical representation of the universe...

-There will always be something missing, because this description should be describing itself.

-It is because we are here, that we cannot logically answer the question asking why are we there.

-You can say it like that, Spyridon. But read my theorems, you will see it is interesting and irrevocable. And then for the "He wouldn't do this to us, up there", if you think about it you will see that it is a good thing. Without the incompleteness, we would have no reason to live, creativity would not exist, proofs could be provided by machines. If you see Albert again, suggest that he reads my theorems.

-Do you consider that mathematics is a branch of physics, or vice versa?

-For me, mathematics "exists" as well as this table or this chair. But obviously we know it only through the physical processes of our brain. In this way, I am a "Platonist".

For me, mathematics describes physical processes so well, because at the end of the day the universe is mathematical.
-But your incompleteness, is this to be "found" in the mathematics, which "exist" or at the level of the physical processes, which enable us to know them?
-In mathematics itself. But you see that all knowledge is preceded by a prioris, mine is to think that mathematics exists.
-The problem I see is that mathematics includes all sorts of constructions, which do not correspond to anything physically, such as spaces with multiple dimensions or strange geometries.
-How do you know, Spyridon, that these mathematical objects do not correspond to anything! Have you already been to the United States?
-Umm! Yes and No.
-Oh! I see. I would not have believed that my question was undecidable …
-I mean that I am a cartoon character and I go everywhere. Why?
-I would like to find some work over there with all these murders going on in Germany … take a little soup …
-I can reassure you. I read that you will stay with Albert actually, at the Institute for Advanced Studies in Princeton, from this year. He will have certainly read your theorems by then.

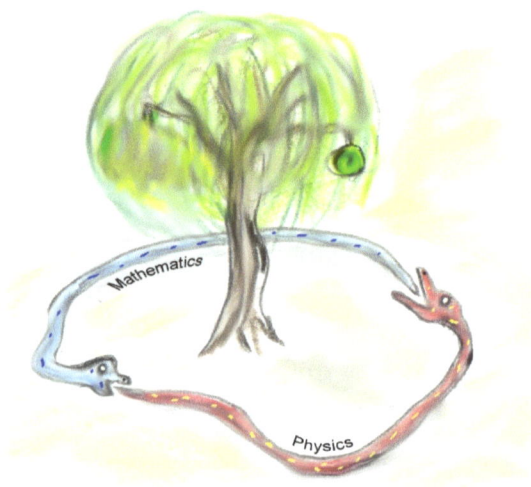

Spyridon comes away from the meeting with Gödel rather exhausted. He doesn't fully realise the enormous consequences of what he has learnt, But it's as if his main reason for being has evaporated. Finished is his dream of ultimate knowledge through the sciences and finished is Einstein's dream of a complete description of reality in a mathematical model. One will never know everything and, worse still, Gödel has probably proved it. As far as Spyridon is concerned, the purpose of his life needs to be reviewed.

He stretches out on a lawn and relaxes.

Tears flow from his eyes. "He made a world for us so designed that we remain in the dark, but just light enough for Gödel to show that we shall stay in the dark for ever" he thinks, "what illusions I've been under, what unbridled pride I've been guilty of, supposing that by work and through logic I shall eventually understand "Him".

"And what lack of humility I display now, refusing to accept reality".

"Rather than being happy at knowing more about it, I am collapsing in the face of what I have just learnt".

In spite of his comforting reasoning, his tears continue to flow.

"I am the proof that even the purest objectives are being pursued for doubtful reasons".

He does not even take his pad out of his pocket. He does not want anything any more. The hours go by. Night falls finally and the sky fills with stars. Slowly, contemplating the stars, Spyridon falls asleep.

11
SPYRIDON MAKES THE POINT

When he wakes up, one of the sentences in his yellow notebook, a sentence to which he promised to return one day, fills his mind:
I am a part of the universe.
"I am a part of the universe, it is the universe which decides, that is to say, which calculates through me. And if it has given me this urge to know, I must follow it and move forward. If it has given me the capacity to attribute meaning, I have to look into it". This reasoning seemed a little bit simplistic to him, not proactive enough, but, for the moment, this is the way he feels, so he is quite relaxed. "It's up to me to interpret Gödel correctly, using other assumptions". And this "end" that he had been feeling before falling asleep, appears to him like a new beginning. Seizing his notebook, he carefully writes down a summary of his thoughts:

- The very process of knowing places a limit on what is knowable.
- One element of the whole cannot describe the whole. As any theory is a product of the universe, no theory can possibly describe it in its entirety. By the same token, mathematics cannot totally prove itself.
- Without indetermination, the world would be rigid, leaving no possibility for evolution and life as we know it. Everything would be accessible by calculation. We would be useless.
- As a matter of fact indetermination is a present from "above". A present which allowed our evolution, a present which has serious consequences, but which makes our role clear to us: to do what machines cannot do, to understand, to make sense of things and reorganize things as a function of this.

Then on another page:

• It is not for me to take everything in hand, I am part of the big calculation of the universe; but by trying to understand, I contribute to it and maybe I accelerate it.
• By "knowing", I handle information and organize it locally, as if this were my contribution to the universe.

He remembers the comments of the priest earlier on and notes:

• He wanted me to understand something, but I wasn't ready. Faith goes beyond the limits of self-reference. Believing is an evolved and personal form of knowing. We are not only capable of understanding and of attributing meanings, but we are also capable of believing and thereby finding answers. Searching, for me, means loving, he also once said, but I don't really understand that yet.

He scribbles furiously:

• By definition, if "I believe" it is that "I do not know". If I knew, I would not need to believe.
• Knowledge is what can be shared with the other thinking beings, faith is personal.
• Faith precedes knowledge, it is something else in another place.
• Knowing and believing are connected, if I could not believe to start with, I would never be able to know. But once I believe, I have to try to know.

Spyridon feels he's coming away from a dead end, he is outdoors once more, in his element, on his lawn among the stars, filled with more energy than ever before and with a greater desire to know than he had ever had. He realizes that at the bottom of his desire to know was an incommunicable faith.

He feels he can now see more clearly, that he knows better where he stands. Discovering means freeing oneself, he thinks. Before the discovery, the spirit is the prisoner of an obsession, like a liar who cannot escape his lies any more. Discovering is a bit like confessing, forgiving oneself. It releases an incredible, vital energy.

However these new revelations give rise to lots of questions. Opening his notepad he comes across his list of 10 questions and his programme of initial work:

1. Why are we here?
2. Who made us?
3. Who made the world?
4. Was there a beginning? And before this beginning?
5. Can one know the laws, which govern the world?
6. Are there extraterrestrials? How can we meet them?
7. Can one meet God?
8. Why does it seem as though I don't understand, while the others do seem to understand?
9. How does one recognize the truth?
10. Is mathematics a branch of physics?

They seem so naïve at the moment and he doesn't believe any more that knowledge consists in formulating then answering questions. Naturally questions are necessary, but they are only a sort of support for the spirit. But he notes on the opposite page:

1. We are capable of attributing meanings, the universe calculates through us. The anthropic principle.
2. Question filled with presumptions, no answer.
3. Idem
4. We cannot know the system from the inside.
5. Not completely.
6. To consider. I think so.
7. I think so, ... I have an idea to try out.
8. They do not understand more than me. I believe that they accept more easily than me in general. It is not evident to go to the heart of things.
9. ???
10. Yes, in one way, no in another...

He crosses off questions 5 and 8. After a moment of hesitation he also crosses off question 10.

"I am not totally satisfied, he says to himself, but I never would be, this is the way things are. Any answer implies presuppositions, which can be always questioned". After this reasoning he also crosses off question 1. "I obviously have no scientific answer, but I have my answer".

"If I really want to understand, I cannot escape it, I shall have to know more about what He has produced, up there".

"His most important creation is the universe, I am going to study it and discover him through his work, as one does with a writer, a painter, a criminal or whomever".

Spyridon goes back to the library, where he had read the Discourse on the Method. He finds out about Ptolemy's Vision of the Universe, but he is especially struck by the destiny of Galileo. How a simple instrument, which increases the scope of our senses, forces us to modify our concepts.

How society resists these new approaches... But what he needs is a more recent input and nothing is better than a conversation face to face.

Spyridon has heard about Carl Sagan, the astrophysician famous for his television shows, who started the SETI programme, listening to extraterrestrials. He decides to visit him straight away. Carl will certainly be able to enlighten him about extraterrestrial life and intelligence, since this is his question 6 and it seems approachable, concrete and less prone to the unconscious hypotheses of others.
He draws a small green man calling Earth.

Hello SETI, can you hear me?

12
IT'S SO BIG, THE UNIVERSE

I want to know the truth, wherever it is. To discover it, we need imagination and scepticism. We shall not be afraid to speculate, but we shall clearly distinguish our speculations from the facts. Carl Sagan

Carl is in Arecibo, Puerto Rico, on this day in March 1990 in one of the radio telescope buildings. When he arrives, Spyridon is impressed by the giant reflector more than 300m in diameter. Carl begins by showing him around the site.
After the tour and over a good cup of American coffee, the discussion starts in earnest.

- Tell me Carl, is the universe really so big?
- Yes, so big that we cannot even imagine it.
- So it's infinite?
- Probably, I'm still not quite sure.
- Well, does it have an end or an edge?
- Not necessarily, imagine a flat, two-dimensional being, who lives on a sphere, on a ball. His world is finite, but without an edge. Our universe has more dimensions, but it's perhaps the same thing. Everything depends on its curve. It could also be that it is infinite, but that all its substance is concentrated in one region.
- But it seems that it has a beginning, despite what Einstein thought.
- Yes, that's what we think, it is about 14 billion years old and it has been expanding ever since.
- And before?

-??? Perhaps there is no before, even your question disappears …..before. Nobody knows. To speak about before the beginning, it would be necessary to be outside the universe. For us who are inside, this is meaningless.

-Speed is the limiting factor. The speed of light limits our possibilities of travelling to other galaxies.

-Yes, the nearer a rocket gets to the speed of light, the more its mass increases and the greater the energy required to continue to accelerate: This is special relativity.

- Yes I know. I know Albert.

- Oh, you know Einstein, but how old are you? He has already been dead for 35 years. The nearest star apart from the sun, Proxima Centauri, is already 4.3 light years away, or 40,000,000,000,000 kilometres. Just to count this number you would need over 35,000 years.

FASTER THAN LIGHT

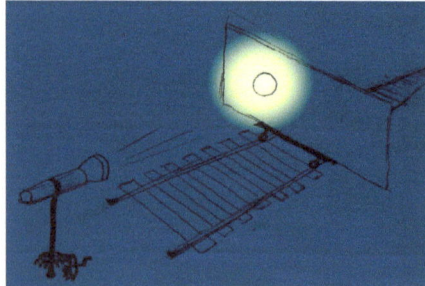

Operation:
By turning the lamp, the spot of light moves on the screen. If this spot is far enough away and the lamp turns fast enough, the spot ends up going faster than the speed of light.

The fallacy:
The spot is not an object or information, which is moving!

- Are we alone in the universe, do you think?

- That would be a tremendous waste of space! I don't think so.

- You mean we are not alone, but there is little chance that we could meet another intelligent civilization.

- Yes, it would take a huge coincidence. In time, nobody knows how long an intelligent civilization can last, and the survival times would have to coincide. In space, distance would need to allow travel, providing there were no short cuts, tunnels linking one region to another, white holes….; meanwhile we are listening out, at least as long as they don't cut our budgets.

- SETI

- You know about the SETI programme? The idea is that an advanced extraterrestrial civilization could be sending intelligent signals that we could recognize…

- Have we heard anything, so far?

- No, but it's like looking for a needle in a haystack, the universe is so vast. Just in our galaxy, the Milky Way, there are more than two hundred billion stars, and there are billions of galaxies. At the present moment, we are practically sure that there are no type 1 civilizations (our level) within 1,000 light years of Earth, because we would have detected them by now, but we are only listening at between 1,000 and 3,000 MHz.

- We must wait. We can only get to know a sphere of 42 billion light years around Earth, the Hubble's bubble, beyond this the photons have not yet had the time to reach us since the big bang 14 billion light years ago. In this sphere of 4×10^{26} metres, we estimate that there are 10^{21} stars……

- But what is there beyond that?

- ????

- Did I say something funny?

- Watch your words, Spyridon. The Hubble's bubble has a horizon. The photons, which come to us from there, are 14 billion years old and were emitted shortly after the big bang. Your question unwillingly contains a temporal element. What is there beyond; when?

Extraterrestrial…

-Oh! I understand. If we mean at the time of the emission of photons, then by definition nothing, because we are at the time of the big bang and because the universe is contracted, but if we mean now … we shall know in 42 billion years. It is like when one sees ships appearing from behind the horizon, there is something further away, but we cannot see it.

-Penzias and Wilson received the Nobel Prize for their discovery in 1963 of fissile radiation, which allowed the rejection of the former theory of the static state, which claimed that the universe was homogeneous in space and time and that it had always been so. Einstein was wrong to want to stabilise everything by introducing in his theory a constant.

-If the universe is dilating Carl, it is dilating in space, or at least in something.

-??? You see how ordinary language is not very appropriate.

-I am stupid. But what are these shortcuts, these white holes that you mentioned earlier?

-At the moment nothing has been proved, but it could be possible that black holes have an "exit", a white hole in another region of space or somewhere else, with the whole constituting a sort of tunnel … At the moment this is science fiction and even if these "worm holes" can theoretically exist, they would have real stability problems.

- So there is little hope of using them for the moment.

- If the universe is infinite, statistically there must exist several replicas of you. Of the Earth and even of our complete Hubble bubble…

The discussion goes on late into that night.
Spyridon leaves Arecibo again, having been able to observe through the instruments. He is daunted by the limitlessness, which he has just glimpsed: billions of galaxies, pulsars, quasars and stars in the various stages of evolution and the black holes, which Hawking had already discussed. A gigantic ballet, way beyond our imagination, but of which we are part. All this has to have a meaning. What a chance to be a cartoon character, who can break through any barrier and meet some marvellous people. What good fortune to be able to take the time to reflect about all of this. What could be more important than looking for its meaning?

Spyridon takes the time to think about what he has just learned. What "He up there" has created is gigantic and so inconceivable, that one cannot help feeling a sense of being small, of admiration, of happyness and of infinite love. "It is miraculous that the universe provided man with a brain, mathematics and the capacity to understand "he says to himself". Miraculous that He gave him this desire for knowledge and a basic means of understanding".
The other aspect of the work is "the infinitesimal aspect" and Spyridon does not want to miss anything.
One day, while going through the list of Nobel prize winners for physics on the Internet, he comes across Murray Gell-Man's name. The inventor of quarks. Murray is described there as one of the most important brains of the 20th century. He is certainly worth talking to. Spyridon draws a Jaguar and he knows from his readings that this will bring him directly to the right person.

13
THE TWO PHYSICS

We are in 1985 when Spyridon arrives at Caltech where Murray is a professor. Without introductions the discussion gets going:

-You see, Spyridon, it is a question of scale. We are a little in between, neither really small, nor really big. This is true for size, but is also true for duration. Who, for instance, can really appreciate a long succession of events, which happen in a billionth of second? Or a cause, whose effects appear only after 5 billion years. Nevertheless, some of these events are essential to an understanding of the universe and have a profound effect on the determination of its destiny.

-So if we "lived" in another scale, we would see things otherwise. We would take other causalities into account.

-In another scale, of space, time, energy, speed, acceleration… Yes, everything would be different for our perception and appreciation of things.

-What can we do to complete our knowledge, if our only possible vision of reality is so narrow?

-Well, we look for laws, which could be valid for all scales…

-And is this the case for the laws of physics?

-Well no, and this is the problem, making equations, which work for quarks, molecules, human beings, black holes and galaxies. Einstein dedicated the latter years of his life to this, without success.

-Yes, I know that he believed in this "unitarian" theory. But what are we lacking according to you? Are we still far off?

-At the moment, it's as if we had two physics. Quantum physics, which works well for small scales and which is probabilistic and without a time arrow; and relativity, which works for big scales and which is deterministic with an elapsing time. We do not yet know how to unite them. Maxwell had already unified electricity and magnetism. We know how to integrate weak and strong nuclear forces into this. But the last stage, that of the integration of gravitation presents an obstacle. Now these four forces were integrated at the beginning of the universe. There was only a single force, which decomposed subsequently. If we want to go back to those first moments, we have to have a "unitarian" theory.

-What hopes do we have?

-Several roads are being pursued, and some seem promising such as the M-Theory. But even though we don't reach the objective fully, we are improving our understanding. A unitarian theory represents the ideal, the important thing is the road that we are following. You should go and ask this young professor Smolin what stage he has reached, he is working hard on this.

-I am trying to understand if there is a will behind this universe. Why do we have the taste for understanding and the talent for mathematics? It surely cannot be all in vain.

-The risk in the unification of physical theories, is that there is not much room left for a will. In a unitarian theory, the "why" and the "how" join together somehow.

-You mean that it could be possible that all this exists without a will or a creator!

The interlocking of scales and virtual worlds

-Maybe, if it were possible to establish a unitarian theory. We could be in an universe for which all the causes are integrated.

-Would He have played such a trick on us "up there", not even existing? He has allowed himself such little room for manoeuvre to intervene or to show himself...

-Oh! We are far from understanding the simplest things. And the things that we believe we understand are often illusory. With some friends last year, we opened an institute in Santa Fé that starts from totally different premises and tries to understand the notions of complexity and emerging properties. You should go and visit, it's fascinating.

-I shall go there.

-Mankind belongs to a whole, the universe and there he occupies a limited space-time (even if this expression is inadequate). He considers himself as a separate entity, but it is only an illusion due to the narrowness of his awareness. All the quanta of matter and energy have interacted at one time or another since the big bang, the universe is a "unique quantum system", which responds in its entirety to all interaction.

-I had already felt this! " I am a part of the universe".

-Yes, there is no isolated "reality", but a tissue of relations …

Spyridon spends the night with Murray talking about quarks, these ultimate constituents of matter, bosons and other elementary particles, components of matter and symmetrical groups. All this vibrates in Spyridon's brain. But apart from the physicist, Murray is master of so many other disciplines.

They discuss living, complex adaptive systems, the development of life, intelligence …

Finally Spyridon outlines to Murray the idea, which is beginning to form in his mind.

As he leaves, he notes in his pad:

"Murray expresses it differently, but also seems to think that the existence of a unitarian theory could imply a rigid world, as I had deduced from my conversation with Gödel. In theory it would be possible to deduce just about everything from physics. We would not need chemistry, biology, sociology... In practice this cannot be achieved. And anyway, to me it seems suspect. For me, there is something else. One does not appreciate a poem by studying the alphabet in greater detail."

"If we succeeded in reuniting the two physics, the initial peculiarity, which makes Einstein's equations fail, would disappear and the question about what preceded the big bang would take on its full relevance again, pushing further back the limitations of self-reference" Spyridon says to himself.

The summer is scorching in Geneva in this year of 2003. The crowd swarms on the banks of Lake Geneva trying to cool down. Physics is not going to teach me much about man, thinks Spyridon. He decides to find out more about the "masters of the planet" and goes to Paris to meet Yves Coppens.

14

MEN, DINOSAURS OR SHARKS ?

P aris in the summer of 2003. Yves Coppens is the happiest man on earth. A man of tremendous natural gentleness and endless patience, his attitude reveals a man concerned with present reality, an optimist and a visionary. He is teaching at the Collège de France. Spyridon, for his part, feels in a provocative mood today.

-All right, the Earth is not the centre of the universe, nor of our galaxy nor even of our solar system, but in the end human beings are still the masters on Earth! Spyridon exclaims provocatively.

-Do you know, Spyridon, when human civilization dates from?

-2,000 years since Jesus Christ and maybe 5,000 years of civilization before that.

-One can say a little more, 40,000 years since the Early Paleolithic age and the invention of fire, 6,000 years since the Neolithic period and the Bronze Age. Dinosaurs for example reigned on Earth for more than 100,000,000 years. An extraterrestrial visitor to the planet would have been far more likely to come across one of them than us… Not to mention certain insects. They have been there for more than half a billion years and they will certainly outlive us, or sharks who have been in existence for 400 million years with virtually no significant evolution.

Certain species last hundreds of millions of years, while others disappear quickly, but no species is eternal. At the level of the planet, it is not which sort dominates at any given moment, which is important. Nature evolves. In some periods, one species appears to dominate and to live all over the planet, then something happens...

-So even on Earth, man is only a small passing phenomenon?

-Yes, I believe this very happily, there is not much chance that we shall last for a long time in the present form, or if so, we are going to need a little more wisdom and a little more humility to know how to protect what we need to survive.

-And we're not doing this?

-Of course not.

-Don't you think that with a bit of progress we could find some way of surviving, which nature had not foreseen.

-Have you been reading " People Magazine" or some kind of pseudo-scientific article?

-??? No, I am serious.

-It is possible. But we are part of nature, we do not "find" anything; possibly nature finds or experiments or happens through us. Us and other species maybe more promising, for example insects…

It is likely that we are sitting on a dead branch, we possess the means of complete self-destruction and the kind of primitive reptilian brain ready to push the button, and the kind of violence which will finish up doing it.

-Oh! The animal in us …

-No, just we ourselves simply. The rest you think you see is only cosmetic. Our hope is thought.

-What are we to do?

-It is important to understand and know about previous history and how things evolved, to get to that stage. This may allow us to be more aware of steps to come. Sometimes the consequences of an event show themselves only much later by jumping millions of years or generations. Murray must have told you, it is a question of the scale of one's observation.

-I understand. Knowing a system or a state at any given moment is not enough, we must understand the logic of the evolution, which led to this state.

-Yes.

As he leaves Yves Coppens, Spyridon reviews what he learnt. "In fact, he thinks, I am in the process of building my own image of "reality", I am starting to "understand" it in my own way, putting some order into the various elements of the puzzle. Since I am aware of the limitations of logical knowledge, I look "up there" and discover His work on various different levels. Everywhere I notice how much I have to remain modest in my approach, however I have the impression I am progressing. At every turn, new questions intrigue me".

"First of all this question of two physics: on the grand scale, gravitation dominates and the universe seems so well-defined that "He up there" has practically excluded himself from any way of intervening. Then on a smaller scale, quantum phenomena dominate, our description is based on probability and things seem to happen in a completely illogical way".
"Then, there's man's place in all this and the mystery of his capacity to know..."
"So many things still escape me, for example how do we connect our knowledge of physics with knowledge acquired in other less fundamental domains such as chemistry or biology..."

"And even with physics, when I read or I hear Murray say that a particle can be in several places at the same time, I cannot really accept it, even though some people explain it by saying that it was nowhere at all, before we observed it. I am a little like Albert, I have a preconceived image of what reality must be, for me, we must be able to understand it, if not globally at least in every stage of our discovery."

Spyridon decides so to go to Denmark to visit the Niels Bohr Institute, a cradle of quantum physics and what Spyridon calls the School of Copenhagen.

In 1913, Bohr published his historic paper on the quantum theory of atomic structures. He was then a professor at the University of Copenhagen.
The institute itself was founded in 1920 and managed by Niels Bohr until his death in 1962. During the War, Bohr was offered a post in England. The allies were afraid his expertise in the matter of nuclear fission might fall into the hands of the Nazis. This period cost theoretical physics heavily, due to the displacement of its main protagonists.
Very many personalities of the world of the atomic physics stayed in Copenhagen and visited the institute between 1920 and 1935.

Spyridon is happy to go and draws a little mermaid ...

Part three: Understanding

15

BEAUTIFUL SOPHIE FROM COPENHAGEN

It is November 18th, 1992, when Spyridon arrives in Copenhagen. The institute has preserved an air of the 20s, of its heyday, although numerous buildings have been added. Old measuring instruments, photos in black and white showing groups of the famous physicists of the time still decorate walls. Naturally Bohr is always in the centre. On certain pictures one can recognise Planck, Einstein, Dirac, Heisenberg, Schrödinger, Pauli, Fermi, Bethe and many others.
Spyridon is expecting a conversation to match the appearance of the premises... A young lady in jeans looking very relaxed enters the room and politely says to him:

-My name is Sophie and I am here to inform you about your future studies at the institute and to fill in some forms with you. You have arrived on the thirtieth anniversary of the death of our founder and I am alone in the institute.

-Oh! Niels Bohr... I only came to ask some questions.

-Of course, all the students do this and I am here to answer you.

-How do you explain that a particle can be in several places at the same time?

-Oh! My friend... Surprising ... It is an old story... For a long time one has not been explaining any more. One observes.

-Oh! How you do you observe such an impossible thing.

-It is from this point that quantum physics was developed, often right here. We are in the cradle of the biggest part of today's physics. Our pragmatic approach has been very successful. Quantum physics is the best theory that mankind ever conceived, the most tested, the most precise. Quantum mechanics predicted antimatter, radioactivity and so nuclear energy.
It has enabled the production of semi-conductors, transistors, processors, explained super-conductivity and led to the invention of the laser and the nuclear-magnetic echo - to quote but a few examples. Practically all our technological objects arise from the quantum theory.

-Yes, but you do not answer me. You say that a particle is in several places at the same time, but did you observe it?

-No, one cannot observe it, because of "decoherence". By observing a particle, one forces it to make a decision and to be in a particular state, a well-defined position.

-If you cannot observe it, how can you know it? What do such terms as "to be" or "to exist" mean for you in these conditions?

-You do not ask the same questions as your companion students Spyronton, I find you a little bit strange, as if you come out of a cartoon.

-Yes, this is indeed who I am, how did you know?

-No, I do not know you. In which section would you like to enrol, Spyricon, so that I can complete the forms.

-No, I simply want to ask you to explain to me how you can assert something which you have no way of observing.

-Oh! Yes. You see Spirliron, it is necessary to distinguish between quantum physics and the interpretation of quantum physics. Here at the School of Copenhagen, we follow the thinking of our illustrious founder, Niels Bohr. For us "doing science" is having equations, which work perfectly and which allow us to anticipate a host of things, as is the case for the Schrödinger equation, the wave function. So, to answer you, "knowledge" is knowing how to handle these equations.

-Then you give up "understanding"?

-Yes, because when one makes a "measurement", an "observation", "decoherence" appears. The equation collapses and the particle is only in a single place. Our intuition is not adapted to these levels, it didn't develop for these analyses, so that "understanding" has no meaning for us. Evolution provided us with intuition adapted to physics, which we need every day and which had a value for the survival of our ancestors. It did not equip us to understand this type of phenomena, and there is nothing amazing in giving up observation. We have to expect that certain things will seem strange to us, don't you agree with me?

-Yes, all right, but this is purely theoretical. What do you call a measurement or an observation? Is it the presence of the measuring instrument or the consciousness of the observer? Is this the reason there are two physics? Is understanding inevitably connected with the intuitions we are used to? Aren't there other ways of understanding? How you can give up so easily and, besides, you still have to convince me you are right to give up!!!

-Oh! Calm down Spyridon, we have not given up anything. On the contrary, we simply don't want to introduce false debates, which all depend on interpretations ... If we had spent all our time in philosophical quarrels, we would not have been able to move forward and achieve the magnificent results, which you know and which have served all mankind! Anyway, it is impossible for us to observe a quantum state, a superposition state at our level! So what could the term "observe" mean at the quantum level! Sometimes pragmatism is also useful in science. What the wave function describes is perfectly established but will escape us for ever. What we do know is that it is the result that counts.

-There is, therefore, a whole hidden world that we can describe with the wave function, but never observe or understand?

-Yes. Obviously, on the question of the interpretation of what this world actually is, I have to admit to you, we do not all agree, even though very few questions the theory itself. Can we enrol you now?

-Well, I am not satisfied at all. I think that the main thing it is to understand. The rest may be carried out with machines and may not be so important... Let us see...

-Then you do not want to register?

-No, no, no, I have no time for it. Can you recommend "something else", Sophie?

-??? You mean some other institutes?

-Scientists who think differently about "understanding".

-Oh! Yes. You could visit David Deutsch in Oxford, he may be closer to your point of view. Or Anton Zeillinger in Vienna.

-I will go!

-But they won't enrol you, you know.

-Too bad for the enrolments. I shall still go there. But tell me again quickly. If particles can be in several places at the same time, then why can't I, or other objects?

-Because of the decoherence, precisely. As soon as a particle becomes bigger, there are parasites and the wave function collapses. There is decoherence. When one changes scale, one's physics changes too!

-You mean that the microscopic behaves in one way and the macroscopic in another. What is true for a group of particles (me for example) is not true for an isolated particle.

-Yes, that's it. You are in an established state, evolving according to determinist laws, in a time which is elapsing. This is not the case for an isolated particle. In its case, there is a fundamental indetermination and time is not elapsing.

-Always?

-Well, not if you take the example of Schrödinger's cat. For it, the indetermination flows back to the macroscopic level.

-Who is this cat?

-It is a cat imagined by this Austrian, Schrödinger, in 1936, to build a bridge between the microscopic and the macroscopic, it is in a black box, closed and perfectly sealed. Inside the box, there is a quantum mechanism which, when activated, is in a superposition state, in several states at the same time. It is arranged that one of these states opens a valve, which releases a mortal gas, which would kill the cat. So long as one does not open the box, the cat is both dead and alive. When one opens it, one forces the system to choose and the cat finds itself in only one of these states. If you register, you will be able to study the cat in depth.

-Schrödinger didn't kill any cats though, did he?!

-No, rest assured, it was purely a mental exercise. Although it was tried recently with other mechanisms replacing the cat and Schrödinger was right.

-Yes, I understand, the cat and the box provide a means of applying quantum indetermination at the macroscopic level. This is really interesting. I wanted to ask you again Sophie …

-Yes, about the registration?

-No, about Descartes.

-Oh! Yes, the poor man! He has suffered as a result, naturally. It is an old story. One can no longer consider the observer on the one hand, and the observed system on the other one. This division between the spirit and the world as the basis of the Cartesian paradigm collapsed in the 1930's. We all have a little difficulty in adjusting to this.

-Thank you for receiving me Sophie. I would indeed stay a little longer, but I have to visit this David Deutsch now. I have to get back my "understanding", it is fundamental for me, you know. I fully understand your idea that observing, or measuring makes no sense at the particle level, but there must be a way to overcome this difficulty and to provide a meaning for the superposition state. I am convinced and I want to follow this right through to the end.

-You do know, Sirondin, that at the quantum level causality does not work any more? Yes. Yes, the experiments of Aspect and Gisin have confirmed this. At the quantum level, time and so causality do not exist. And yet you claim to understand, Spitilon, that establishing causal relationships is essential and you still want to go somewhere else.

Schrödinger's cat in the box

- Maybe time and causality do not exist because of your interpretation of these experiments, I have to go down my path Sophie.

-Are you leaving Sputucon? Will you come back to the institute? Take a registration form with you, one never knows … I will give you my e-mail address …

Spyridon is very impressed by the beautiful Sophie. He would have willingly stayed, he would have even signed the registration form. Sophie's arguments were making him doubt. But, contrary to millions of others, this is not his destiny, at least in this universe. How, wonders Spyridon, if time does not exist at quantum level, how is it able to manifest at our level?

For the moment, he cannot wait to visit David Deutsch. The idea that it is necessary to give up understanding upsets him. How had all these great minds been able to give up understanding? He was certain that there had to be another approach and he was looking forward to hearing about it. There also had to be a way of getting round decoherence. Otherwise it might overthrow his ideas completely.

As he leaves the institute he stops for a moment in the school's library to find out more about Schrödinger, the author of this famous equation, the wave function. This reassures him. Schrödinger had not accepted Niels Bohr's position either...

However Spyridon is satisfied. The gulf between two physics is now clearly defined, he even begins to feel how things are connected. Two ideas come to mind: information and emerging properties.

He makes a drawing of a sort of particle and adds a rocket borrowed from a colleague to indicate that he is really in a hurry to get to Oxford.

16

PARALLEL UNIVERSES

Noon has just sounded on this June 10th, 2000, as Spyridon knocks at the door of David Deutsch's house in Oxford. Nothing happens for a good while. He rings again, it is only after the fifth ring that he hears steps tumbling down the stairs. The door opens, revealing a David, with bristly hair.

-Did I wake you, David?

-No, no, I was only sleeping lightly.

-Who are you? What are you doing here so early in the morning?

-I am Spyridon and I came to ask you some questions.

-Ah! Spyridon, Oh! Yes, I was expecting you, you were to come this morning. Come in and don't look at the shambles.

-Is it true that a particle can be in several places at the same time? Do you think as well that we have to give up understanding and only rely on equations?

-No, no, understanding is essential, because understanding enables one to ask questions. Without understanding we would not know which question to ask. A good theory should enlighten our understanding, widen our point of view, allow us to ask new questions, build a bridge between two areas which seemed different, explain inexplicable things, simplify our vision. In brief without understanding, we would not be doing science but a sort of botany of the 19th century, counting, classifying, but not progressing. Understanding is the motor, this is what drives our intuition and our spirit, classifications and calculations, experimenting comes later. When we have a theory that works and has been verified, but which we don't understand, we have to work at it's interpretation, otherwise it will only deliver half of its contents to us. All the equations on special relativity by Einstein were already well-known, by Poincaré for example, before Einstein's publication. His genius was to provide a frame of interpretation, an "understanding".

-So how are we to understand these particles, which are in several places at the same time?

-Yes, well no. Who said that they were in several places at the same time? It is not that. Everett already produced his thesis on a surprising interpretation in 1957, entitled the Multiverses. Listen!

-I'm listening

-There's not only one universe, but billions and billions. And every particle has billions of replicas in every universe. There are billions of Spyridons, who come waking up billions of Davids.

-Ah!

-If you take a dice with six faces and you roll it, you have one chance in 6 that the 3, for example, comes out. In fact, the six faces all come out, but in different branches of the multiverse. Naturally you only see one, because your consciousness is only aware of one of the universes. But I have to say to you that they don't all agree.

-Who do you mean?

-The physicists of course. But I am going to prove it in an unmistakable way.

-How you are going to prove that there are billions of universes, to which we have no access, where billions of Davids exist, to whom you will never speak!

-It is very simple and I am at the end of my demonstration.

-Oh?

-Yes, I just need a quantum computer to be built.

-What's that?

-It is the device, which exploits the properties of quantum physics to handle information in a way, which would be impossible on a computer. You see, instead of working with only 0 and 1, quantum bits called Qubits can take all the intermediate states corresponding to their doubles in the other universes, the famous superposition state, billions of operations happen in parallel. This will accelerate the calculation enormously. I've done for quantum computers the same work Alan Turing did in 1936 with his machine, for conventional computers; I have described the precise functioning of a universal quantum computer and proved that such a machine could exist.

-Yes, but how does this prove that there are billions of universes?

--A quantum computer will be able to carry out operations, for which, even by taking the whole universe as its memory, there would not be enough capacity to store the results of all the intermediate calculations.

-For example?

-Multiplying two numbers, even with hundreds or thousands of digits does not really pose problems for a computer. By increasing the number of digits, one increases the calculation time, but only in a reasonable way, about 31% for one figure more. With the inverse operation, factorisation, the determination of the prime factors of a number it is more complicated. The calculation time on a conventional computer increases exponentially (with today's algorithms), so that by adding one figure, the calculation time triples. For example, the factorisation of a number of 250 digits will require 10^{500} simultaneous calculations. There are about 10^{80} atoms in our Hubble's bubble, so one can calculate its maximum "memory" and it is not enough. One will be forced to admit then that Everett was right. A quantum computer shares information with an enormous quantity of replicas of itself throughout the Multiverse.

-!! Incredible, totally incredible!!

-With the concept of the Multiverse, everything becomes more simple to understand and to describe. Decoherence sheds its shroud of mystery and is only the result of our limitation of being aware of only one universe. Time itself has a simple explanation: instead of being a means to describe change, the passage of one state into another state, in the multiverse, time becomes a way for our consciousness to establish order among all the possible states. The problem of the creation and the beginning of time is thus obviated.

-??? Incredible, totally incredible!!

-The universes, which compose the multiverse are static, change is only an illusion, which results from the fact that our brain establishes an order in its successive awarenesses of universes.

-And do these billions of Davids in these billions of universes do exactly what you do?

-No, not necessarily. This creates new universes. Every time I make a decision, there are some Davids who decide the opposite and go off into new circumstances. That's why everything is determined at the level of the multiverse, but we are free all the same in the universe where we are conscious. A global physical theory has to be a theory based on the multiverse!!

-How do you mean?

-Everything possible exists in different junctions. And new junctions are developing non-stop. We are conscious in one universe only, this is why we have a choice. At the level of the multiverse, choice and indetermination disappear. The quantum computer will give us irrefutable proof of this.

-But does this quantum computer exist really?

-Yes, the first ones have been built and in 20 years we shall have such powerful ones, that they will be able to break any code and carry out calculations, which will always be beyond the reach of conventional computers.

-Do you have one?

-Oh, not me, I'm only a theorist. But go and see Isaac Chuang at IBM.

-And doesn't the "global physical theory" you were speaking about contradict Gödel?

-No, not within the framework of the multiverse...

The discussion with David goes on all day and through the night. Spyridon has never met another man with such a broad vision and such a lively intelligence. David takes scientific knowledge seriously and pushes every theory to its extremes.

David is not limited with his knowledge in quantum physics or information theory. He thinks in a global way by taking into account all human knowledge. He builds bridges between apparently unconnected domains and makes everything so transparent.

David made a remark when they discussed the theorem of Gödel. Without Gödel, in other words if there were completeness at the level of a universe, it would have been possible to build a computer, which could prove the truth of a proposal, by a simple check of the series of symbols. There would no longer be any need for mathematicians or for "understanding".

Understanding and creating become therefore the essential contributions of mankind to the multiverse.

Without non-determination, any development can occur mechanically. Worse still, our discoveries can only become more and more superficial as time goes by. Thanks to Gödel, we know that there will never be a fixed method of determining if a proposition is true, or a fixed method of generating mathematical or scientific knowledge. For this, man is necessary. Spyridon is thoroughly convinced of it. He has heard it three times and now with three different arguments.

"Man and maybe the quantum computer…", he thought.

Quantum computers, cryptography and the multiverse.

Peter Schor of Bell Laboratories discovered a simple algorithm in 1994, requiring only a simple hardware compared with a universal quantum computer. This algorithm enabled the decomposition of numbers into prime factors. This factorisation is a so-called "non tractable" operation. This means that on a classic computer, the calculation time increases exponentially in relation to the number of digits making up the number being factorised. It has been estimated that a number with 250 digits would take billions of years to factorise, even linking all the computers in parallel that existed in the world in 2003. A quantum computer applying Schor's algorithm could do it in a few seconds by using the interference of 10^{500} universes. It is this enormous quantity of universes involved in the operation that makes the calculation manageable on a quantum computer. One could imagine the consequences for cryptography, whose current processes are based on the problems of factorisation.

It is also interesting to notice that the number of intermediate calculations stored during factorisation is greater than the number of particles in the classic universe, which enables Deutsch to claim that this is proof of the existence of the multiverse.

The main argument against the multiverse is Ockham's razor, propounded by the English logician of 14th century. Ockham's principle is the economy of means: it is useless to suppose the existence of something, which one cannot observe. Give up everything that you can give up.

Why would nature produce such a complex structure. According to Deutsch, the multiverse is much more simple than one of its branches, the conventional universe. It does not require that one specifies initial conditions or physical constants, it does not need the collapse of the function of the wave.

17
A LIBRARY CONTAINING ALL BOOKS PAST AND FUTURE

" A bookseller in dark glasses asks him: " What are you looking for? ' Hladik answered: ' I am looking for God. ' The bookseller says to him: ' God is in one of the letters on one of the pages in one of the four hundred thousand volumes of Clementine. My fathers and the fathers of my fathers looked for this letter; I have become blind from it. " J.L. Borges

Spyridon has wanted to meet Borges for a long time, as he has read some of his works. The occasion presents itself on this July 19th, 1984 in Geneva, two years before his death. Borges is almost blind. They walk along the harbour and finally settle down on a bench to discuss.

-Apparently you have a library that contains all the books that have ever been written in the world, says Spyridon.

-Yes, that's true. But I also have all the books that will ever be written in the future if you want, and many other less interesting things.

-What do you mean, can you show me around?

-Naturally.

-Where is it? Spyridon wonders.

-In my brand new Apple computer.

-What do you mean?

-If you give the order to this machine to print all the possible combinations of let us say 5 million characters, it will come up with a huge number of books, each with a different combination of the 5 million letters.

-So far, so good, but what's the point?

-Well, you will have all the books you want … Among the books you will have Shakespeare's complete works.

-All right, says Spyridon after a moment of reflection.

-You will have all the books that man has ever written and all the ones he is going to write.

-Yes, provided that these are books of less than 5 million characters, it is true that you can have them in several volumes. But what's the point?

-Well, you're looking for answers to your questions and they are certainly in these books.

-All right, but it is going to take me a long time to find what I am looking for and even if I find it, how can I be sure that I have found the right book and not a book written by a trick of the combinations. Your library looks like what Internet will become (you don't know it) in a few years. It is so polluted with waste, that it's no good any more.

-You are right, everything is there, but nothing makes any sense. If you "understand" what you are looking for, you're the one who knows or doesn't know how to select it. The information alone is not enough. Didn't I hear you were a firm supporter of "understanding"? Information without understanding is not much use. Unless you can make a programme a bit more subtle than the one that prints out combinations.

-Ah! Of course, I can introduce rules of syntax, which would already eliminate a good part of the rubbish. Then, I can put a dictionary and other grammatical rules into it… In fact, if I want the books to look different from something the monkeys typed, I'd have to introduce in the program some laws of physics. And also mathematics, naturally. But I would also have to put in some biology, otherwise how would the programme evaluate my sense of colours or my feelings …

-In fact, you should put all the laws, which govern the reality we know, into this programme.

-Yes, that's it. With this enormous programme your library would become more useful. But that comes back down to observing the information, which is in nature, and to know how to make something of it, I would need a model of it, and therefore to know the laws and so be able to "understand"…. In fact, your programme would be that of a perfect universal virtual reality machine.

-Yes, I agree.

-In short, you can see that it is we who attribute meaning to information, and if there is no structure to our thinking, no understanding, there can be no attribution of meaning. And the more one's thought structure is refined, the more reaction the attributed meaning will allow. But if your programme could be really complete, then meaning would be lost again!

-What do you mean?

-Well yes, there would be no more decision-making possible, because everything would be already written, you could read tomorrow's newspapers and those of a thousand years' time... You see one is clearly talking about a world which is not ours..
-I understand. In fact, it is Gödel's incompleteness that you are getting back to: the programme must be incomplete. In our reality, there cannot be a complete programme to generate the library's series, you have just demonstrated this again...
-What is interesting, it is that the library is eminently compressible, because all you need is one line of programme to publish it. But by lengthening your programme, you decrease the number of printed bits of information, you make it less compressible, but you "increase the meaning", up to a certain limit, that of Gödel.

This gives Spyridon a lot to think about. The connections between information, compressibility, meaning and incompleteness are beginning to unfold. He is realising more and more the central importance of the notion of information connected to the notion of meaning. All he has ever heard about or learned is information.
He has always concerned himself with the attribution of meaning to information and it is certainly this, which is his role in the universe. However he still has to understand the changes in scales.
One day while he is absorbed in his reading of Kolmogorov, his publisher telephones him to say he is invited to go to the Santa Fe Institute.

18
AT THE INSTITUTE OF COMPLEXITY

Santa Fe is a private research body situated in New Mexico. It is a multidisciplinary institute confronting scientific horizons as different as physics, biology, data processing and social sciences. Since its foundation in 1984, Santa Fé has two main features: an interdisciplinary approach and its studies of complexity and complex adaptative systems..

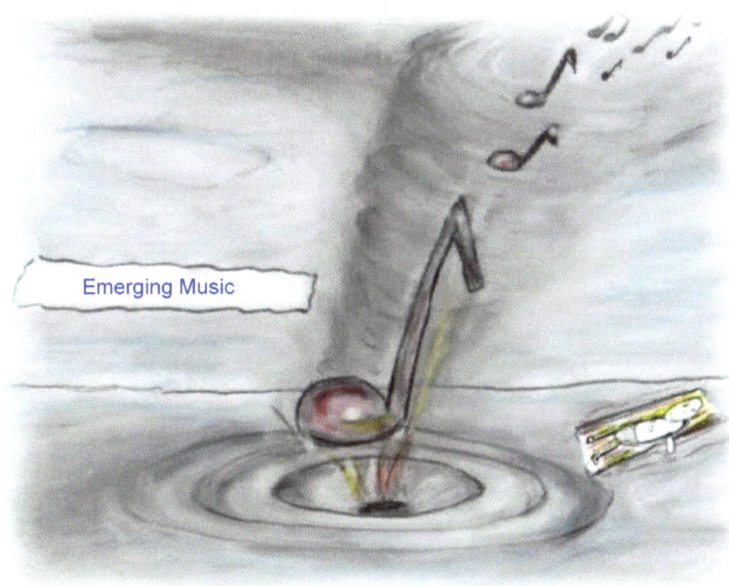
Emerging Music

-Certain structures of the universe become more complex in time, now Maxwell taught me long time ago that with time, there is a dispersal of information and an increase in entropy.

-You mean Carnot. But you are right, the second principle of thermodynamics applies only to isolated systems. If you consider a system in which organization increases, your brain for example, it is not isolated, all sorts of exchanges are going on with the outside world. If you managed to isolate it, entropy would increase and information would dissipate bit by bit, or even brutally for example, if you stopped its supply of oxygen to the blood.

-But how has the universe created entities as complex as the human brain?

-I do not know, but one notices this "complexification" everywhere, even if we still have to define this term. The universe creates complexity and, above a certain level of complexity, properties appear, which we call emerging, properties of the group, which are not the simple sum of their constituent properties. Maybe consciousness and understanding are emerging properties of the brain, due to the incredible quantity of interactions between the billions of neurons of which it is composed.

-Are computers going to become conscious, and understand?

-Perhaps when computers have reached a certain level of complexity, a sort of consciousness will appear …

-What will they think?

-Nobody knows. Emerging properties are one of the big mysteries of the universe… We try to explain why certain systems tend to become complex and organized. Why does an organization suddenly appear and order establish itself where chaos should normally reign? But we're only at the beginning.

- I do not understand Chaos, I believe it is only our model or interpretation who is unapropriate. Can complexity appear from very simple initial data?

-Naturally. For example the representations of fractals seem extremely complex to us, but they are generated by the repetition or iteration of extremely simple equations. Iteration, this is the application of the same formula to the result of the previous application, can produce surprising results. Regularities can appear at unexpected moments.

-If I were a person born in your beautiful fractal and if I investigated the universe around me, it would seem extremely complex to me and I do not see how I would have a chance of finding the simple equation, which lies behind it all.

-This is the difficulty of research. Here is another example you can play with: Take an ant, which is walking on a grid of black or white squares, like a stone floor. To start with, you can suppose that all the squares are white. The ant moves away from the central square let us say northwards. It moves off one square and observes the colour of the square on which it then arrives. If it arrives on a white square it paints it black and turns 90 degrees to the right. If it arrives in a black square, it paints it white and turns 90 degrees to the left. Then it continues to apply these rules every time. These very simple rules produce complex behaviour. During the first 500 steps, the ant returns non-stop towards the central square. During the following ten thousand steps, the picture it leaves becomes very chaotic. And suddenly, as if it had decided what it should do, it repeats sequences of 104 steps, which take it two cells towards the south and it continues with regularity in this way, forming a beautiful band, as straight as a highway. After ten thousand steps, order eventually emerges from a situation which seemed chaotic... In order to see this, we have to change perspective...

Complex ants with draught-board

-I see, simple rules can produce an apparently very complex behaviour and conversely, very simple behaviour can emerge from apparently chaotic behaviour.
-Yes, that's it, this is the kind of thing that we study here. You should come to our summer seminars, especially you who are interested in self-reference.
-You're not trying to enrol me, are you?
-Uhh, no …
-Oh good, what is the connection between complexity, emerging properties and self-reference?
-If you look closely at all the appearances of complexity, which we envisaged, there is a form of self-reference being applied.
-Interesting. Really. Ah! Yes... Thought is naturally always about itself, we only know how to think about other thoughts, with the ant and the fractals, there is iteration... This question would require more thought. But I sense something behind this. The universe, if I consider it as a gigantic computer calculating its following "state" from its "previous state", is also a gigantic, self-referring, iterative system, oh gosh...
-It's up to you to think about it, Spyridon, but for my part, the most surprising thing is the link between systems, which together produce emerging properties.
-Eh?

-Fossil remains indicate that the composition of the atmosphere and the temperature of the surface of the Earth seem to have been continuously regulated by the living. Although the complex network of retroactions, which maintain conditions suitable for the continuation of life is largely unknown, it is certain that the complete biotope is responsible for it.

-How do you mean?

-Take, for example, the stabilization of the level of oxygen in the atmosphere. This percentage (about 21 percent) has been maintained as such for millions of years. If this concentration increases by only a small percentage, this causes the spontaneous combustion of living organisms. If it falls, aerobic bodies die from asphyxiation. Furthermore, without this emerging property of the biotope as a whole, chemical analysis suggests that all the other gases, which react violently with oxygen, must have practically disappeared. Now, hydrogen is 10 billion times too abundant, carbon dioxide 10 times too plentiful, compared with what they should be, if one takes only the chemical interactions into account without any mediation at all.

-Yes, I see.

-The global luminosity of the sun, meaning the total energy emitted by our star, has increased more than 50 percent during the last 4 billion years. Now, according to the fossil analyses, the global temperature of the Earth has remained strikingly stable around 22 degrees centigrade.

-You mean that the regulation of the global temperature of the Earth is also an emerging property of the whole biotope.

-Yes, although we still do not understand the mechanisms. We must consider levels of organizations with emerging properties at every level. From the quark to the universe. From writing to stock-exchange phenomena ... That is why Santa Fé is really interdisciplinary ...

-I understand, but I would like to ask you another question. When you read a poem, you sense a meaning, you understand something, which goes well beyond the actual words and sentences, is this also an emerging property of the combination of reader and writing? Or, if I ask the same question in another context: would it be correct to say that physics will never describe nature completely, as long as it does not integrate the emerging properties, which are studied by the other sciences?

-I believe, Spyridon, there are levels of organization with emerging properties at each level. One day maybe a physics including emerging properties will integrate chemistry, then neurology, psychology... Meanwhile, we really have to understand how these emerging phenomena work, we are right at the beginning. Today if we wanted to give a physical description of a chemical reaction, not only would it be too complex, but we would lose a large part of its "meaning". The reason we can study chemistry, biology and other sciences is the emerging phenomenon. High-level phenomena are comprehensible, but they cannot be deduced from low-level phenomena. "You came with a yellow pad, because you thought you would be taking notes" is an explanation that one cannot deduce from low-level phenomena like the study of particles. It is related to emerging phenomena such as thought and contributes a "meaning" which would escape us if we were relying on a purely "physical" study.

Interlocking of scales
and virtualities at the time
of the dinosaurs

-I see. The "sense" or the significance, which we need in order to be able to "understand" results from our capacity to work on the emerging phenomena and to create new ones ourselves!

-Tell me, could "time" as we know it at our level be an emerging phenomenon, since it doesn't exist at the quantum level?

-It is very possible, we don't understand yet.

Spyridon feels transported with joy, a missing link is now within his reach. Everything is coming into place: complexity, information, emerging properties, self-reference, meaning, understanding, incompleteness - how can we group these all together? His brain is bubbling, he thinks of the road he has so far travelled. Out of curiosity, Spyridon takes another look at his old list of 10 questions:

"One asks questions and naturally one expects to find answers without realizing that one is getting into a much more complex system, where everything is connected, where emerging phenomena and self-references become blurred, and very often an answer to one question would represent the answer to all the questions. And the answer to all the questions would mean the non-existence of the universe such as we know it. The question - answer arrangement can be only a guide, true answers are beyond answers, they emerge in us".

However, against question 9 he records: it is necessary to go out. And against question 1 he notes: We are not there, we think there...

"Which property would emerge from a quantum computer specially programmed to simulate an "exit" from the universe?". He needs to think and sort out his plan.

"It's incredible" he says to himself, "how long it takes to see clearly. There are too many new concepts whirling around inside me". He decides to return to Cambridge and the peace of the gardens and the green lawns and the calm atmosphere of the university will help him straighten things out.

He chooses a beautiful summer day and finds himself sitting on the grass on the banks of the Cam.

Spyridon thinks again about the pledge he has made: to devote his life to understanding. How had this universe, of which he is part, been able to produce this desire in him, what was the purpose?

Understanding is an emerging phenomenon of a higher level. The universe has developed endless complexities, including life itself, to reach the desire to understand. From living has emerged awareness and then understanding with self-reference as the main characteristic. With awareness, of course, comes the awareness of being aware.

Universe, complexity, life, emerging, understanding, self-reference. Spyridon juggles with these words again, trying to put them in order, to situate them alongside each other. Each time he comes up against an obstacle. Suddenly an idea comes to him. An idea that Maxwell had evoked at the time, funny he had not paid much attention to it. Understanding is creating virtualities. The universe had "willed", in a way, this kind of "virtual reality machine" that we call awareness.

Spyridon remembers his conversation with Borges. It's by "handling" virtual reality and improving the programme that generates this that we know how to impact on reality. Physics, as a virtual reality programme, has enabled us to develop technologies. In fact, reality and virtual reality are complementary in a way. The one causes the other to "emerge", which then in turn impacts back on the other. The universe is acting through me thinks Spyridon once again: "I want to know more about virtual realities".

The moment has come to visit his friend Philippe Rochat, an eminent computer scientist. He draws a good plate of Italian spaghetti and the job is done.

19
IT WOULD ONLY BE A GAME

Philippe Rochat is not only a big gamester, but also an eminent programmer. What's more, he lives close by and is a cartoon fanatic. Spyridon walks across the campus of the Federal Polytechnical School of Lausanne towards the data base laboratory. Philippe welcomes him with his usual smile, he has so many things to show him and so much to discuss, but to begin with he sets him up in front of a simulator.

-One could imagine oneself aboard a real plane.

-"Real", which reality do you mean by that? I'm joking, the rendering is almost perfect, the only thing I don't know how to recreate is naturally the sensation of acceleration.

-You would need a bigger computer?

-No, that wouldn't make any difference. Whatever computer you have, the only way of creating acceleration, it is to accelerate the player or then to rummage about in his brain to stimulate the impression of acceleration.

-I understand.

-All the rest is only a question of the processing speed, or slowing down the rhythm of the brain to leave the calculation time to happen.

-You mean that I can always know if I am in reality or in a virtual reality, by observing how my body behaves with regard to acceleration.

-Free-fall poses the same problem. In the simulator you are isolated from everything. But one can not isolate you from the ground gravitation to simulate free-fall or a flight in space.

The Rochat Simulator

-Except of course, if you knew how to rummage around in my brain to give me the sensation of floating.

-Yes, of course.

-In fact you answered one of my questions. Without rummaging in the brain, I can always know if I am in "reality" or in a virtual reality.

-Let us imagine that without knowing it you are a permanent prisoner in a virtual reality machine programmed with false laws of physics. For you, your world is the one that you have always seen, that of the machine. Do you believe, Spyridon, that you would be able to tell that you were in a virtual reality?

-Let us see, I would try to know how the world works, I would discover its laws bit by bit (that is, the programme which operates the machine prison). I would try "to understand", to make forecasts. I would not be content with knowing the laws of this universe, I would like to know its origins and the attributes of the various constituents of this world.

-All right, you would write treaties on physics, mathematics, biology….

-It will be necessary for these laws to be rather complex to explain my presence in this world, meaning that this world should be sufficiently complex to allow emerging properties, awareness….

-The programmers who built the machine maybe anticipated all this.

-The programmers had a limited time to conceive the machine and cannot intervene any more in the system, unless they produce what I would consider from the inside, as a "miracle", whereas I have all the time I need. So, one day I would end up discovering that something "is not working", that there are contradictions in the system: either at the level of the origins that do not exist, because everything was programmed from the outside, or at the level of emerging properties, I would always be missing something. I would eventually conclude that there must be an outside world in which my world was built. So that this does not happen to me, the programmers who built this world would have had to make it fully self-explanatory, meaning fully self-contained in terms of explanations. Now, since no part of reality can have this property, I would deduce that there is an exterior. Maybe I would need millions of years, but if I am curious and persevering, I would be bound to get there.

-But if you came up against an indetermination in our "reality" here, or found it impossible to get back to the origins, or the need to refer to the outside, what would you think?

-It would need to understand what was behind this indetermination, hem, in fact I don't know, I would have a strong feeling that there is something outside…. interesting, Philippe, interesting...

-Are you thinking of Gödel?

-Of course, but also about our universe, and the multiverse and especially about Him up there, as Albert would have said.

-I have never met him! But do you know that your own imagination is also a virtual reality machine?

-How do you mean?

-When you imagine or think, you reconstitute a sort of reality with its own physical and logical laws, which more or less well "portrays" outside reality... When you dream for example, your programme does not include the usual laws of nature, you fly, you fall from 100 metres without injuring yourself. Once awake, your programme is more complete because other parts of your brain are functioning.

-All right, our brain constitutes a more or less good simulation of "reality". Ah! You mean that knowledge is always virtual?

-Yes, knowledge is a virtual emerging phenomenon about "reality", which is possibly also virtual.

-You speak about "virtualities" wrapped inside each other like Russian dolls, which can never be a perfect "representation" of each other, because of Gödel, although they are emerging from each other.

-Yes, but interlinked in multiple dimensions...

-Philippe, could you, if you had powerful enough computers, programme for me the simulation of an entry into a black hole?

-I don't know those things well enough, it would be gigantic, you would have to simulate gravitation, the whole of physics would be involved..., it is practically impossible, you would have to integrate all the emerging phenomena, what happens exactly inside the black hole...., my wife has prepared some spaghetti ... Let us see...

"I presumed that if my reality had been created from the outside, so, if it is a virtual reality, I am bound to find this out eventually, if I have enough time. Because at a given moment I shall stumble across contradictions. Now, does the incompleteness of my current reality imply that it is virtual? "

"Didn't "He up there" have an infinity to programme our universe? Or is there a whole chain of "Them up there"? Like "up there" Russian dolls, fathers and sons, all virtual, together forming an infinite chain, in which contradictions and indetermination disappear... Is this the multiverse? "

Spyridon thinks for a long time and bit by bit his plan becomes clearer.

" Without realizing it, this is this plan that I have been following since the beginning".

Everything is still milling around in his head: self-reference, emerging properties, quantum indetermination, the theorems of Gödel, multiverse, compressibility, virtual realities, God the creator. They all begin to interlink, he "understands", yes, Spyridon understands, he understands the reason for his own existence.

What immense joy, what tremendous satisfaction! What peace of mind again. "Thank you..."

"In the meantime, let us be concrete, he says to himself, I need to know more about the famous black holes, if I want to be able to help Philippe".

It is the moment to meet Lee Smolin, Murray had recommended to him. He knows that this meeting is important for his plan. But he wants to be ready. What he still needs is to verify his ideas on information as the corner stone of his construction. Jacob Bekenstein is the man who is going to be able to help him.

A small drawing and here he is in Princeton in the 1970's.

20
EVERYTHING IS MERELY INFORMATION

Jacob is studying at Princeton in 1976. He is a pupil of the famous John Archibald Wheeler. Spyridon strolls among brick buildings of this prestigious university for a while, where so many people and famous thinkers have walked. It is in the famous library, where Gödel and Einstein once took their afternoon tea together, that Spyridon meets Jacob.

-Space and absolute, eternal time, in which matter moves and interactions take place, this is a prehistoric concept. That must date from Newton.

-So what?

-Everything is discrete, Spyridon, there is a smaller energy, a smaller length, a smaller time, Planck went down this route. Achilles and his tortoise did too, but they were precursors.

-You studied the maximum information contained in any region in space and showed that it is limited!

-Yes with Stephen Hawking, while studying black holes. More exactly, all the information contained in a space in a three dimensional area can be written on its surface. It is proportional to the entropy of the surface and this is why everything is discrete. There is an absolute limit to the information, which can be contained in any region. It means that by definition space is discrete.

-???

-The holographic principle of Gérard Hooft claims that we are wrong to think that the world consists of "things" occupying regions in space. Damn' Newton! The only things that exist are "screens" on which the world is represented.

-???

-Yes, only screens, borders if you want, which can exchange information with other parts of the universe.

-Do I understand that the substance of the universe is neither space, nor time, nor matter or interactions, but information itself.

-Rather simplified, that's right.

-Like a vast library. Or a gigantic computer.

-Space would be one description of the channels, which allow information to circulate. And the geometry of space is a measure of the capacity of the screens for relaying information. In brief, according to the holographical principle, the world is only a network of relations, and the history of the universe a stream of information.

-This seems to agree with what I was thinking a long time ago, lying in my hammock. The universe is only information as a gigantic hologram produced by the interference of billions of the other universes. And information is precisely the only thing that we know how to handle, this is what produces emerging properties.

-As every screen is limited in the quantity of information, which it can contain, there is no global or central vision of the universe, but instead a different universe for every screen.

-Yes, every region contains limited and different information. The study of the universe, is therefore in a sense a branch of the theory of information.

-Yes, Spyridon. But more than that, there is no global vision of the universe. This enables us to answer questions like what is the maximum information we can stock in a given volume:

In a sphere of one centimetre diameter, one could store 10^{66} bits. The visible universe contains around 10^{100} bits of entropy, which in theory could be stored in a sphere with a diameter of one tenth of a light year. These figures are well beyond the storage capacity of our current technologies. We can however now describe the characteristics of the "universe computer": To simulate the universe in all its details over 14 billion years, a computer would need to store some 10^{90} bits and would need to perform 10^{120} manipulations with these bits.

-Our whole reality would then be the calculation by a huge quantum computer determining its own future all the time.

-This is roughly what Seth Lloyd says, professor at MIT and former colleague of Murray. The elementary operations of the universe consist in moving particles and making them interact.

-Would its basic software be what we call the laws of the physics? Physics would be only a branch of the theory of the information?!

-Yes.

-Then 2400 years later, Platon and Aristotle would reconcile themselves, as they are both right!

Spyridon writes in his notebook:

Hologram, interference, multiverse, information, emerging properties, quantum computer, simulation, black hole, He up above, interlinked virtualities.

This is his plan and it is now almost ripe. Spyridon has decided to organise it. But who will be able to help him finalise his project?

His hand draws without his even noticing it.

The dye is cast, he will go and see Lee Smolin.

21

SPYRIDON'S PLAN TAKES SHAPE

Spyridon only needs a short journey through time to 2001 to find himself very near the place where he met Bekenstein, but 30 years later. Lee Smolin seemed at once terribly nice with his black beard and big glasses. Spyridon introduces himself, he uses some of Murray's words. But Lee was clearly prepared for his visit and, without even saying hello, jumped straight into the discussion. How can he know why I have come to see him, wonders Spyridon, but Lee's attitude puts him at his ease.

-I know your plan Spyridon, the technical problems to achieve it seem to me to be insurmountable, but if you cross the horizon of the black hole, no force will be able to stop you falling into it and coming back, in a sense, you will be outside the universe and basking in information, which is not accessible to us. Is that what you are looking for?

-Yes, but before that, tell me: are there many black holes? How do I choose the one I need?

-Probably there is a huge amount, at least 1 billion billion within our Hubble's range. I would choose a very big one to minimize the tidal effects, and not too far away, so that we can collect the maximum of information. Cyggnus 2 would be a candidate.

-And you think that certain regions inside the black hole generate new universes.

-Yes, I claim that the physical laws of the new universes are linked to those of the mother universe and that the whole system of universes operates by a sort of "natural selection". The universes that allow complexity, evolution and emerging properties reproduce with more efficiency than the others. You see, the fairly big and complex universes will allow the formation of stars and their evolution in black holes and will naturally generate a greater number of baby universes, because they will have more black holes.

-Are there equations that are already sufficiently refined to be used to programme a simulator?

-What are you trying to say? Ah I understand, you want to simulate a journey across the horizon of a black hole rather than make the real journey. It would be less risky!

-No, simulating the journey, means making it! In a universe made of bits, physics becomes identified with the simulation of physics. If the universe is only a bath of information, a hologram, a perfect simulation capable of restoring a black hole on a quantum computer …could generate a "real" new universe. This new universe would be only a part of the gigantic calculation.

-What you say deserves thinking about. It reminds me about what John Archibald Wheeler was already claiming in the 1980's. He considered that all particles, all forces and space-time itself derived their existence from binary choices. And it is true that we now know how to make computers with almost any material. Matter calculates. However, Spyridon, we are a long way from all this. We lack concepts, equations, mathematical methods and of course powerful enough computers... And even if we had all these one day, who knows what would be the result. We would still need hundreds of years …

-I have time, everything will sort itself out. And especially "I believe" that "I must" do it, it is my role, says Spyridon, measuring his words carefully. A quantum computer using the resources of billions of universes, a simulation so perfect...

Lee and Spyridon are to spend several weeks together. Conversations with Lee make everything clearer and his help is effective, Spyridon's madness seems contagious.

The yellow pad gradually fills up with plans, equations and tasks to be carried out.

Finally, by the time Spyridon and Lee part company, their ideas are precise and the plan is ready. Spyridon will follow his destiny. He smiles to himself: "To know more, there are no other solutions, one must believe more..." Supreme paradox. But supreme logic: "before calculating, you must know what to calculate!"

There is another small leg however.

Spyridon makes a small drawing of a quantum computer …

22

THE QUANTUM COMPUTER

January, 2002. The IBM Almaden research laboratory in San Jose to the South of San Francisco is as secure as a safe. Obviously, Spyridon has not got the necessary documents and papers. It's no use arguing with IBM's security. As he is about to give up, he has an idea, maybe a small drawing could bring him directly into the laboratory of Isaac L. Chuang. On his yellow pad, or rather once yellow, so much had the colour faded, he draws the face of the physicist. Chuang is now in front of him smiling.

-I had nevertheless sent you an invitation Spyridon.

-Where did you send it?

-To your publishing house, of course.

-Ah! I never go there alive and when I come away from there, I am already stuck on the paper.

-Yes I know, you are a cartoon person, David warned me.

-It makes no difference ...

-You wanted to see our 7 Qubits machine didn't you? You may be disappointed, there is not much to see, it is a simple molecule in a test tube with some equipment all around to send and collect information.

-No, tell me just how it works roughly.

-We use quantum properties, atoms and molecule nuclei cut to size as the processor and memory of the computer. In fact 10^{18} molecules in a superposition state, are involved. The big problem is the insulation to avoid the decoherence. You know the problem of the Cat.

-When will you be able to you process Schor's algorithm on one of these machines? And when do you foresee a universal quantum computer?

-Schor's algorithm is already working. On our 7 Qubits machine: we have got as far as factorising the number 15. The machine has found the two prime factors 3 and 5. We are very proud of it. But to factorise big numbers, we shall need several thousand Qubits and several more years of work. As for a universal quantum computer, even though it would be possible in theory, we are it still a long way off. If ever we get there one day, different technologies are possible, but we have no idea which will be more promising ...

-You look very sceptical!

-Yes, but I am working on it here and then remember in 1949 when the ENIAC (Electronic Numerical Integrator and Calculator) was equipped with 18'000 vacuum tubes and weighed 30 tons, Popular Mechanics claimed that in the future computers would probably only need 1'000 tubes and weigh only 1,5 tons. Schor has already surprised us with his system of error correction. At first, we never thought of being able to extract the results from the system and of avoiding decoherence. Nevertheless we moved on. Here, read the article which has just appeared about us:

> SAN JOSE, Calif. December 19, 2001 - scientists of IBM in Almaden Research Center have achieved the most complicated quantum calculation ever made. They forced a billion billion molecules specially conceived for the purpose to become a quantum computer of 7 Qubits in a test tube. They thus resolved a simple version of the mathematical problem at the heart of the security of coded data.
>
> " This result reinforces the idea that one day quantum computers will be able to resolve problems, which even the most powerful of today's computers, working for billions of years could not resolve", reported Nabil Amer, manager and strategist at IBM …

Spyridon explains his project to Chuang and informs him of his discussions with Lee Smolin, David Deutsch and his other partners in this adventure. Chuang appears totally sceptical at first. Several conversations later, though, he is finally convinced. After all it is only a cartoon. It won't cost IBM a penny.

One year later, Chuang, Smolin, Rochat and the others have established an excellent working relationship. The project is progressing quickly. Everybody agrees that it will only take another 200 - 300 years to complete it.

23
READY FOR THE GREAT DEPARTURE

Spyridon feels ready for the last great journey. He believes in his plan, although it is not actually his, as it belongs to the universe. He feels peaceful, almost distant, the best brains of the planet are at work, a job that will last 300 years and in which he will not participate...

This last journey will be virtual, in a simulator powered by a quantum computer networking with millions of other computers and specially programmed "to render" his entry into the black hole.

Spyridon will join the finished machine in the future. The chosen date is 1st January, 2300.

His friend Chuang put the quantum computer into production in coordination with David, and Philippe is taking care of the programming and networking, Lee is coordinating with the programmers to make sure that the most recent mathematical models are being used. The whole team is at work and each member has the job to appoint his successor in due course, because, unlike cartoon characters, people eventually pass away.

Everything must be ready when he arrives in 2300.

Spyridon makes a drawing and meets himself at the machine on 1st January, 2300.

Obviously he has been expected for a very long time, and everything is ready as planned.

Each of his friends has left him a small goodbye message, which Philippe hands to him in an envelope.

-I knew your grandfather well. He was called Philippe like you, Mr Rochat.

-I am going to explain how the simulator works and introduce you to the virtual crew. All the latest mathematical models have been installed. The computer can use the resources of billions of universes and it has almost unlimited reserves of energy. It will draw them from the black hole it has simulated itself.

-Incredible! And the principles of thermodynamics then?!

-Simulation generates a real "new universe"...

-In relation to which, we are "above"?

-Yes, one could say so. We worked a lot on the insulation and we understand decoherence perfectly.

-But what happens with the approach of singularity, I suppose, you still have no unitarian theory?

-Yes and no. It is not "complete", Spyridon. However our quantum gravitation avoided singularities. We pushed "incompleteness" back one more stage. One doesn't know exactly how the machine is going to react, it is not really a machine any more. One even knows that it is not possible to totally anticipate it, which is the point of the whole experiment. In principle, everything will be controlled from the screens you see here.
So that we shall always be able to interrupt the simulation if necessary or produce a small "miracle" at your instigation in the machine, the new universe should not escape us. As for crossing the horizon...

-And the creation of the gravitational field at the approach of the black hole? How have you simulated this ...

-You will fully become a part of the machine, you and your atoms will be part of the computer itself, calculations will be made in your own brain, you will be able to work with 10^{46} bits... You'll feel practically nothing.

-Then, I can leave now?

-Before you go, we're going to make a complete back-up, so that we can eventually restore any parts of you which may get damaged during the trip. This is normal before quantum simulations. Are you sure you would prefer to travel in your own body, rather than a "downloaded" copy?

-Yes. Could I stop the experiment, once it's underway?

-You will be able to do whatever you wish, but it won't occur to you stop the experiment, because nothing will indicate to you that it is indeed an experiment. Unless you find out for yourself, which seems improbable to me, the simulation is perfect. Except for incompleteness! But if you had the time and the inclination to push your thought-process to the point of finding out that you are in a simulation, your only possible exit point would still be through a black hole. This would mean that your mission would be achieved anyhow.

-Ah yes. This is what they used to call destiny in my day: being led to choose what must happen.

-Yes, these are geodesic trajectories in the multiverse, those which the majority of your replicas will choose to follow…

-Would you prefer that we start by simulating the simulation?

-No, Philippe, that wouldn't serve any useful purpose. I must leave.

-Well then, everything is ready. We have even prepared a good reserve of virtual barley beer on board, brewed to an eighteenth century recipe. You will have all the comfort you could possibly want, we've taken care of your future needs…

Spyridon hands to Philippe his tattered, once yellow pad filled with notes, as a souvenir, as he won't need it any more, and he goes over to the compression door.

He knows that by going into this machine he will lose the main part of his current identity and most of his recollections. He will acquire new replacements at once. Practically all of his vital functions will be taken over. His consciousness will be changed but preserved, the operating speed of his brain will be adapted to the calculating power of the computer. Everything will take place fluently and anything, which could indicate to him that he is in a simulator and not in his previous reality will be erased. He will become an integral part of the machine and will be the first being in the machine to be conscious of himself.

Spyridon does not regret anything. He has nothing to regret. He was never part of this world, which he is now leaving. He sees so clearly "why". He has his answers, he is his answers, he has performed his role, he has accomplished his faith. He can leave for the next stage, rediscover his origins. Researching is loving.

"It is not my plan, he says to himself once again, it belongs to the multiverse. I am following my geodesic trajectory, whether I like it or not, guided by love".

He approaches the compression hatch and goes into the machine.

24

SPYRIDON PLUNGES INTO THE BLACK HOLE

As he enters the simulator, Spyridon immediately loses any sense of being in a machine. Not a single neuron in his brain has retained the slightest memory of another reality, other than the one he has now plunged into, everything is directly taken over by the quantum computer, of which Spyridon is an integral part. Each of his atoms has been transformed into a part of the computer. After some seconds of adaptation, he remembers very well the long journey towards Cyggnus 2 and clearly makes out the gigantic black hole's disc of accretion. He remembers the preparation and the departure of this space machine to carry out this extremely important mission. The nearness of the black hole thrills him, he is arriving at the end of his quest and he knows it.
In the nearby control room, Philippe and his assistants watch the instruments.

-Fasten your seatbelt, Spyridon!

-And we shan't be ripped apart, Captain Haddock?

-No, damn' it, our machine and its interior are protected by a system of gravitational waves. And this black hole is big enough so tidal effects are reduced.

-Are we still far way?

-We are now approaching Cyggnus 2's horizon. We are going to hover in order to make the measurements and to create our field of entry.

-Be careful, as soon as the engines reignite we shall be plunged into a gas of hot photons!

-Ready?

Some moments later...

-Here we are, the gravitational opening has occurred, we are going to go down and we will lose all contact with the satellite relay.

-Yes, ...

-We are crossing the horizon.

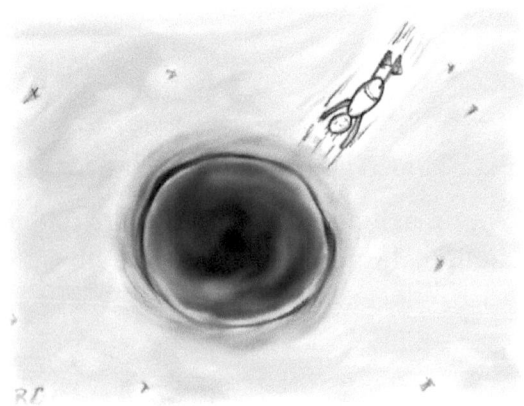

Spyridon plunges into the black hole

As the horizon is crossed, screens in the control room go blank indicating nothing any more. No further information appears. Nothing.

Philippe, in front of the instruments, does not really expect this.

What has happened?

Must he stop the experiment? Nothing comes through from the adjoining room, as if it did not exist any more.

Philippe is worried.

Standing in front of the simulator's compression door, he hesitates to enter a moment, then says to himself that it would indeed be better to give some good thought to it first.

He glances at the instruments again, and at the screens that have turned dark blue. On the main screen appears:

"MEMORY FAILURE. UNABLE TO REBOOT".

What's going on over there, on the other side?

Philippe constructs some hypotheses.

- If inside the simulation chamber it is still in a state of quantum superposition, it would be better not to open it...

- And if the programme was sufficiently perfect and that I've "really" created a new universe. What does "really" mean in this context? In any case, this is beyond me and I have no further "room for manoeuvre" to produce a "miracle"!

-Will the information contained in the machine be irretrievably lost?

All these thoughts crowd into Philippe's head. He opens the yellow notepad that Spyridon had left him, and quietly reads from the first to the last page.

Elsewhere

A churchman approaches the young lad and asks him:

" What are you doing with that yellow notepad, my boy? "

- I was meditating, father, love and faith are our only guides towards the knowledge of God...

25
END

For those who would claim that this story is not correct, Spyridon would have said this:

-If you don't believe that the search for truth and knowledge lies in the heart of every child, perhaps you have a poor memory.

-If you do not think that one can travel through time and meet famous people, then try reading some books.

-If you think that "understanding" is pointless and does not produce any concrete results, look at the human creations around you, because they have all come from the same place, the human brain.

-If you think that understanding is not one of life's purposes, you are missing out on one of the rare things, which has given a meaning to the lives of millions of us.

-If you are among those who want to impose their faith or their beliefs on others, it is first of all because you believe in your faith. Doubtless others do too. It is a good reason to leave them in peace. Faith is personal.

-We are all part of the universe. Nobody owns the truth. And it is better this way.

The "theories" evoked in this story are freely inspired by a certain number of works including:
David Deutsch: The Fabric of Reality, 2002, Penguin
Lee Smolin: Three Roads to Quantum Gravity, 2001, Basic Books
Roger Penrose: Shadows of the Mind, 1994, Oxford University Press
Kurt Gödel: Collected Works, vols 1 to 3, 1985, Oxford University Press
Murray Gell-Man: The Quark and the Jaguar, 1994, W. H. Freeman and Company, New York,
Stephen W. Hawking: A Brief History of Time, Bantam Press, 1988

Short Guide to Science and Fiction

"Spyridon's Quest" mixes science, fiction and speculation, without trying to be exhaustive, I feel I should provide some clarification for the non-scientists, about some of the concepts evoked:

1.- *Travelling through time.* From the point of view of general relativity, time travel is possible in various ways, the story evokes Gödel's solution for it. More exactly, the equations of Einstein predict that journeys into the past are possible near large masses in rotation such as black holes. From the logical point of view, only travel into the future does not necessarily imply paradoxes and special relativity already provides one way of travelling. It is enough to move in a loop at close to the speed of light to find oneself arriving at any time in the future. The choice of the date of arrival depends only on the speed and on the duration of the journey. It is therefore interesting to study journeys into past. In the story, Maxwell speaks about famous paradoxes associated with these journeys into the past causing breaks in causality. Deutsch in his book "The Fabric of Reality" dedicates a chapter to temporal journeys and concludes that only the multiverse allows journeys in the past without raising any logical paradox, the activation of a paradox engendering a new branch of the multiverse. Journeys in time such as Spyridon makes them are anyway totally fictitious.

2.- *The incompleteness of arithmetic.* Few scientific subjects have given as much work to the philosophers and to the thinkers in all areas as the theorems of Gödel. The largest part of these reflections seem to me to be based on unfair extrapolations, although they are intellectually interesting. Since Gödel, many generalizations of his theorems have been demonstrated with, in particular, applications to algorithmic complexity theory. In chapter 19 incompleteness invites Spyridon to believe that our reality may be virtual. As far as what concerns me, these reflections made a broad contribution to the development of my vision of reality. If you research Gödel on the Internet, you will find thousands of references.

3.- *The possibility of a unitarian theory.* The search for a unitarian theory has been one of the big subjects of theoretical physics for almost 80 years. The story raises the M-theory, which is a recent candidate. Lee Smolin's book "Three Roads to Quantum Gravity" discusses the latest approaches.

Stephen Hawking in his best seller "A brief history of time" predicted that we were not very far from it. He has apparently changed his mind. In his article " Gödel and the End of Physics " (that you will find on the Net), he observes that such a physical theory is self-referring as is the theorem of Gödel and that any physical theory express itself with mathematics containing arithmetic and that this suggests that the M-theory will be incomplete. Spyridon gains the conviction about this necessary incompleteness of any global theory by various means, both logical and physical, with the last argument being the one that is put in Bekenstein's mouth following on the holographic principle: " There is no such thing as a global vision of the universe". By combining incompleteness with the other reflections, he deduces two consequences, which are "philosophical shortcuts": God can only be outside the universe and our reality is virtual. From then on, Spyridon goes looking for a means to "get out". (See point 9 below)

4.-Black holes. Here is a subject, which has been widely publicized especially since Wheeler gave them this intriguing name. Einstein whose equations predicted their possible existence did not believe in them and considered that these solutions were mathematical and without physical reality. Nowadays the great majority of the scientists believe in black holes although we have only indirect observations through the study of "double stars", where one of the twins is invisible, but exercises its gravitational force on the other. The universe is probably very full of such monsters and there is it probably one at the centre of every galaxy. Future satellites studying gamma radiation will enable us to examine the question more closely. Today, black holes are much more than the exotic phenomena of the theoretical physicists, they have become a real reservoir for testing theories. Stephen Hawking has demonstrated that black holes are not really black. They radiate their energy progressively and evaporate. At the end there would be nothing left. But if energy can evaporate, information cannot, it should therefore be wiped out by the disappearance of the black hole, and that poses a problem, because quantum information cannot be destroyed! One solution proposed for this dilemma is to consider that the universe cannot last long enough for the black holes to have the time to evaporate. The universe would then be "closed" as Sagan suggests in the story, and would implode at the end of a period of time. In this case, it should have to end in 10^{64} years.

5.-*Complexity and emergence*. Although still badly defined, the concept of complexity is one of the richest to have been introduced during the last 50 years. The Santa Fé Institute in New Mexico was founded to study complexity, because it does not fall into the usual scientific categories. The idea of emerging property seemed to me to provide an excellent basis for reflection and development. In the story, it helps Spyridon better describe the term "to understand" and the way in which we attribute meaning. It especially helps him to understand that complete knowledge will never be contained in a basic theory. We need sciences of a more elevated level not only due to the difficulty of calculation, but also because our universe is made in such a way, that it is perpetually generating emerging phenomena, while getting more complex. When he elaborates his final plan, he seems to think that his quantum simulator will appear in a new universe, which is almost pure speculation! (Although? See point 9 below). I can only recommend warmly Murray Gell-Man's book: "The Quark and the Jaguar", which details these concepts with the brilliance of the founder of Santa Fé.

6.-*Multiverses and interpretations of quantum physics*: Spyridon is absolutely not satisfied by the Copenhagen interpretation, which summarizes quantum mechanics at a formal level. Bohm's interpretation, which foresees the existence of a guiding wave for every particle is, it seems to me, disproved by the experiments of Aspect and Gisin. This leaves Zeilinger's vision, which says to us that what we know is not reality but our image of the reality. I know that I do not do his thinking justice by being so brief. The Internet will inform you in detail. This vision is extremely interesting and would have been able to suit the rest of the story. I chose however to privilege Everett's multiverses, because of its surprising nature, of the renewal in its popularity since the publication of Deutsch's book and of my own conviction. It is in fact surprising that no new branches are really created, contrary to what is related in the story, for reasons of simplification. In the multiverse concept, time does not elapse, at each moment we are aware of one universe and one alone. Spyridon asked Sophie at what moment does decoherence happen, in other words, at what moment does a particle, which finds itself in several states come down to one single state. She replies that it's at the moment one observes it.

This has given rise to numerous scientific and philosophical debates. Some asserted that it was a specific property of our consciousness, which produced decoherence. As long as we do not observe the particle, it behaves like a wave, but as soon as we observe it, it behaves like a particle! The multiverse completely simplifies the question. In Everett's scenario, the state of the universe is described as a wave function, which evolves in a deterministic way. There is no collapse during its measurement, but many tiny superpositions corresponding to a superposition of a complete branch of the multiverse. The multiverse leaves two open questions: if the multiverse contains so many superpositions, why can't we see or sense them? What physical mechanism makes the multiverse appear to us like a universe, where everything has its place? The answer to the first question came in 1970, when Dieter Zeh showed that Schrödinger's equation itself contained the effects of "censure". The states of superposition persist only as long as they are isolated from the rest of the world. Decoherence is the answer to the second question. The classic states (in which every object is in only one definite place) are simply the most resistant to decoherence. In the story, Deutsch asserts that the proof of the existence of the multiverse will be effectively given when quantum computers are able to do operations requiring an intermediate memory superior to the total memory of the universe and Bekenstein shows Spyridon that the complete universe contains 10^{100} bits. The existence of the multiverse is proved mathematically by the fact that the evolution of the wave function is "unitarian". All the experiments have shown until now that this was the case.

7.-Quantum computers. Internet abounds with information on this subject, which seems to have become highly fashionable. If Moore's law (every 18 months the capacity of computers doubles) continues, we shall be confronted in a few years with physical problems of dimensions. Some claim that Moore's law will remain valid because in the meantime we shall have finalized the quantum computer. Who knows. Moore's law measures only the ingenuity of man, it is not a law of nature! It is possible that the quantum computer will only have a few applications, like cryptography (Schor's algorithm), research using gigantic data bases or on the Internet (Grover's algorithm) and of course quantum physics itself.

Richard Feynmann was the first in 1982 to imagine that a quantum system could simulate another quantum system. Deutsch showed in 1985 that it was in theory possible to make a universal quantum computer and described its operation. Various technologies are candidates. The main problem is decoherence, as shown in the story. Deutsch claims that the existence of a quantum computer will show the existence of the multiverse...

8.-Space and time are discrete. Like energy they come in quanta, there is a minimum length, Planck's length (10^{-35}m) and so a minimum time (10^{-43}sec) and continuity has no physical existence. This hypothesis, evoked in the story during the meeting with Bekenstein seems to me to be quite widely accepted. Nature seems to us to be quantified exactly, because information is quantified. The idea of a discrete space is quite old, Bekenstein refers to the paradoxes of the Greek philosopher Zénon, who showed that movement is impossible if one considers space to be continuous! In quantum physics, when one decreases distances in an arbitrary way, infinities appear in the equations. Richard Feynman obtained his Nobel prize for his process of "renormalisation" by "compensating" for them with a mathematical device. This process seems a little bit artificial. A discrete space does not need renormalisation. Certainly, the idea of discrete space or time is contrary to our intuition. As far as we are concerned, we can always put another point between any two points. However, Planck's time and length are so small that we shouldn't be surprised that they are outside our experience.

9.-The universe as a quantum computer. This hypothesis is attributed to Seth Lloyd of MIT and one of the main designers of quantum computers. This vision is also supported by the holographic principle, it resolves a number of paradoxes in quantum physics, such as what Spyridon calls "a hidden world which we can describe with the wave function, but never observe." This hidden world is the calculation, which is made on waves, but whose result we can only see in the shape of the particle. The analogy with the functioning of the computer (where we see only the result) is strong to say the least. You will find other analogies in Lloyd's article, which appeared in the journal "Nature" and in the article by Ross Rhodes: "A cybernetic interpretation of Quantum mechanics", which is on the Internet.

Let us imagine, like Max Tegmark of the IAS at Princeton, a hypothetical universe, bigger than ours, which contains a computer powerfull enough to simulate the complete evolution of our universe. Let's call this simulation T. The inhabitants of T would perceive their world as being as real as we perceive ours. Nothing in present-day physics, Tegmark informs us, says that the universe cannot be matched by as fine as we could wish by a finite and discrete model. If the comments that Spyridon exchanged with Bekenstein in the story are correct, namely that space and time are discrete, we, the inhabitants of this universe T, could never prove that this universe is a simulation. In the story, in discussion with Philippe, Spyridon claims if he had enough time, he would end up discovering incomplete aspects in the programmes of any simulated universe. By virtue of this, he deduces that these incompletenesses or errors will make him think that the universe could have been created from the outside. This is a "philosophical" suspicion, it is not a proof. Let us come back to the simulated T universe of Tegmark. There is a way to avoid practically any calculation for the rendering of T, simply by adding a dimension, so that time does elapse, or by considering that T is a multiverse where, there too, all the times exist each in a different branch, as Deutsch explains. As for the programme which generates T, it will be established from the laws of physics. Tegmark asserts that the algorithmic information of this programme would be so weak, that it would all fit onto one CD-ROM. He also points out that there would be absolutely no need for a highly powerful computer, because everything in T would work at the calculation rate of the computer, including the human brains, which would be there. Tegmark pushes his reasoning further: in order for this universe to "exist", is this CD-ROM even necessary? If this magic CD-ROM could be contained in the simulated universe T itself, the latter would "support" its own existence recursively. In the story, Philippe says to Spyridon that after a certain moment, the simulator will draw its energy from the simulated universe itself. Naturally, it is only risky speculation, but the situation is similar to Tegmark's argument. The first reaction would be to say to oneself that we are in a self-referring situation of the "chicken and the egg" kind, wanting to know which "existed" first, the universe or the CD-ROM. But let us not forget that we put ourselves in a situation where time does not elapse, a universe in 4 dimensions or a multiverse.

In this case, the notion of "creation" or anteriority does not exist. The problem of the anteriority of the CD-ROM or the universe is not relevant. Developments of this kind of argument are to be found with Frank J. Tipler of Tulane University in New Orleans and former pupil of Wheeler, especially in his book "Physics of Immortality", to which Deutsch dedicates a chapter in "The Fabric of Reality".

Spyridon goes into a simulator recreating a T-type universe and, "for the record" notes that even if he could be aware of it, he would willingly do what was in the programme for him to do. In order to notice this, he would proceed as he had already indicated in his discussion with Philippe in chapter 19, which brings us back to looking for the CD-ROM in T! A wonderful, self-referring loop....

10.-Can Spyridon be an integral part of a quantum computer? Seth Lloyd would say yes, and I quote: "If you undergo an MRI (magnetic resonance imprint) of your brain at your local hospital, you activate an enormous quantity of nuclear spins. This is exactly what one is talking about when one makes a quantum computer. It seems that at that moment the brain does not use these resources. However, if we knew how to use them, you would have in your brain the calculation power of billions and billions of computers." (quoted from Edge on the Internet). Naturally this is fiction for the moment, but at least it is not impossible. Tipler would certainly approve of these ideas.

11.-What happens after Spyridon crosses the horizon? Having thought a lot about it, I said to myself that this is for you to imagine...

Science by virtue of its demands: confrontation with experience, Ockham's principle, logical coherence and "objectivity", should have a limited reach. It seems more and more, however, in spite of its "rigour", that it provides evidence of an imagination and an inventiveness, which one has a job to find elsewhere. It is worth looking into a bit, it could help us in more important areas than simply supplying us with technologies, it could provide us with reasons for our existence.

In order of appearance:

Spyridon (1949-2300)	1	Yves Coppens	63
The churchman	10	Niels Bohr (1885-1962)	67
The professor	12	Max Planck (1858-1947)	69
The man without name	14	Sophie	69
Plato (427-347 AC)	15	Edwin Schrödinger (1887-1961)	69
Aristotle (384-322 AC)	15	David Deutsch	72
René Descartes (1596-1650)	16	Anton Zeilinger	72
Gottfried W. Leibnitz (1646-1716)	17	The cat	73
Kurt Gödel (1906-1978)	17	Alain Aspect	74
David Hilbert (1862-1943)	18	Nicolas Gisin	74
Isaac Newton (1643-1727)	21	Henri Poincarré (1854-1912)	77
Edmond Halley (1646-1742)	21	Hugh Everett (1930-1998)	77
Albert Einstein (1879-1955)	24	Alan Turing (1912-1954)	78
Him, up there (?-?)	26	Isacc Chuang	80
James C. Maxwell (1831-1879)	26	Peter Schor	82
Stephen Hawking	28	William Ockham (1285-1389)	82
Jacob Bekenstein	29	Jorge Luis Borges (1899-1986)	83
Sadi Carnot (1796-1832)	34	William Shakespeare	84
Ludwig Boltzman (1844-1906)	35	A. Kolmogorov (1903-1987)	86
The snake	40	The man from Santa Fé	88
Moritz Schlick (1882-1936)	44	Philippe Rochat	94
Adolf Hitler	45	John Archibald Wheeler	100
Bertrand Russell (1872-1970)	45	Gérard Hooft	101
Georg Cantor (1845-1918)	45	Seth Lloyd	102
Alfred N. Whitehead (1861-1947)	48	Nabil Amer	108
Claudius Ptolemy (87-150)	53	David Bohm (1917-1994)	121
Carl Sagan (1934-1996)	54	Gordon E. Moore	122
The extraterrestrial	54	Grover	122
Edwin Hubble (1889-1953)	57	Richard Feynman	123
Arno Penzias and Bob Wilson	58	Max Tegmark	124
Murray Gell-Man	59	Franz J. Tipler	125
The Jaguar	59		
Lee Smolin	61		

Glossary

Anthropic principle: 30,31,39,41,53
Believing and knowing: 51,63,105
Big Bang: 31,57,59,63,64
Black hole: 27,61,99,101,102, 110,113,120
Classification of the sciences: 65,92
Complex adaptive systems:65,87
Complexity: 62,87-94,104,119,121
Compressibility: 86,99
Decoherence: 70-73,75,79,107,110,121,122,123
Destiny: 111
Einstein's relativity: 24-27,32,43, 57,61,77,119
Emerging properties: 62,75,88,90-93,97,99,101-104,121
ENIAC: 107
Entropy: 31,35,88,101,103
Expansion universe: 26,55
Extraterrestrials: 17,54
Factorisation: 79,82
Fissile radiation: 60
Fractal: 89
Gödel's theorems:17,18,42-45,49.50. 63,80,81,86,97-1000,119,120
Gravitation: 13,20,22,26,29, 30,61,67,98,110,113,114,120
Him (He), up there, above:26,46,47, 50,53,59,62,67,97,99,102,111
Holographic principle: 101
Hubble's bubble: 57,58,79,104
Incompleteness, undecidability:15,17, 45-48,86,93,99,110,111,119,120,124
Indetermination:27,51,73,74,80,97,99
Information:35-39,51,75-81,85-88,93, 99-106,115,120-125

Interpretation of Copenhagen:70-75,121
Mathematics:14-19,27, 42-52,59,61,85,96,120
Maxwell's devil: 34
Meaning: 24,30,36-39,50-53,59, 71,74,85,86,91-93,97,117,121
Multiverse:77-83,97,99,102, 111,112,119,121,122,125
Ockham's razor: 82,125
Platonicism: 15,17,19,38,47,102
Quantum computer: 78-82,93, 102-109,113,122-125
Quantum Physics: 24,61,67, 70,71,79,81,121-123
Schrödinger's Equation:69,71,73, 74,75,122
Seti: 54
Schor's Algorithm: 82,107
Self-reference:41,42,45-47,51, 63,90-94,99,120,124,125
Speed of light: 56
Superposition state:72-74, 78,107,115,122
Time: 16,26-28,32,35-39,56-62, 69-75,77,79,97,101-105,119-124
Thermodynamics: 34,88,110
Truth: 15-19,22,39,43,46,53,81
Turing's machine: 82,124,125
Understanding, knowing: 35,36,38,39,41,71,74,77,123
Unitarian Theory: 26,30,32,46, 61-63,110,119,122
Virtual reality: 18,34,35,85,94-96, 98,99,102,109,110,119,120

www.ingramcontent.com/pod-product-compliance
Lightning Source LLC
Chambersburg PA
CBHW041059180526
45172CB00001B/26